Origin and Evolution of Planetary and Satellite Systems

Origin and Evolution of Planetary and Satellite Systems

Edited by
Diedrich Möhlmann and Heinz Stiller

With 61 Figures and 39 Tables

Akademie-Verlag Berlin
1989

Authors:

Siegfried Franck,	Zentralinstitut für Physik der Erde der AdW der DDR, Potsdam
Rudolf Gottfried,	Dresden
Diedrich Möhlmann,	Institut für Kosmosforschung der AdW der DDR, Berlin
Ingo Orgzall,	Zentralinstitut für Physik der Erde der AdW der DDR, Potsdam
Ulrich Schmit,	Institut für Kosmosforschung der AdW der DDR, Berlin
Frank Spahn,	Institut für Kosmosforschung der AdW der DDR, Berlin
Hanno Sponholz,	Institut für Kosmosforschung der AdW der DDR, Berlin
Heinz Stiller,	Akademie der Wissenschaften der DDR, Berlin
Richard Wäsch,	Institut für Kosmosforschung der AdW der DDR, Berlin

ISBN 3-05-500278-4

Erschienen im Akademie-Verlag Berlin, Leipziger Straße 3—4, Berlin, DDR-1086

Lizenznummer: 202 · 100/441/88
Printed in the German Democratic Republic
Gesamtherstellung: VEB Druckhaus „Maxim Gorki", Altenburg, 7400
Lektor: Dipl.-Met. Heide Deutscher
LSV 1495
Bestellnummer: 7636735 (9050)
05200

Preface

Origin and evolution of the Solar system are even for a long time a subject of intensive scientific investigations, but up to now there exists no generally accepted scenario for these evolutionary processes. This is mainly due to an unsufficient knowledge of the detailed properties of the planetary system and early preplanetary and protostellar processes.

Space research gives the hope that as well the properties of the planctary system (by direct methods) as relevant astrophysical processes (by methods of extraterrestrial astronomy) will become observable more in detail. This is the reason for a renaissance of planetogony in our days.

It is the aim of this book to discuss in some detail an approach to planetogony which seems to be a serious alternative to the Schmidt-Safronov scenario of planetary growth by (hierarchical) accretional growth from planetesimals in a low-mass disk. The disadvantages of models of stochastic accretional growth are the long time scales of this growth, the difficulties with gaseous content and connected growth of giant planets, and the impossibility to understand comparable structures in the four satellite systems of Sun, Jupiter, Saturn and Uranus. The proposed alternative approach to explain these comparable structures by comparable gravitational instabilities in disks around a central body is able to avoid these difficulties.

Here it should be noted, that aspects of this alternative approach have been introduced into the scientific discussion by Alfvén, who advocated the role of plasma-processes for planetogony and the importance of comparable properties of the four systems, by Polyatchenko, Fridman, Toomre and Hunter, who investigated the gravitational instabilities of disks, by Cameron with his models, favouring the origin of planets from an earlier "protoplanetary" state (of giant protoplanets), which evolved in a more massive early preplanetary disk, and by Morfill, Völk and Tscharnuter who assumed that planetogonic processes were effective even when the central body was yet forming.

The planetogonic scenario, proposed with this book is based on the assumption of an early massive disk around the yet forming central body. Radial gravitational instabilities are able to form under these conditions. The resulting radial structures are analytically described within the frame of linear perturbation theory. On the basis of the derived structures it is possible to understand the observed radial spacing of orbits in the planetary system and the satellite systems of Jupiter, Saturn and Uranus. The derived densities for the disks indicate indeed the existence of an early disk with a mass comparable to that of the yet growing central body. Additionally an outer thin disk appeared in the Solar system as a predecessor of giant planets. Some of the processes, relevant for the proposed planetogonic scenario are discussed within this book. It is the hope of the authors, that this book will induce increased discussion and a resulting broader consensus on origin and evolution of our planetary system and of our home-planet.

The authors wish to express their thanks to Mrs. I. ZINGEL for her help in preparing this book, and to Prof. Dr. K.-H. SCHMIDT, Prof. Dr. F. KRAUSE, and Dr. sc. G. RÜDIGER for their helpfull scientific comments and discussions.

D. MÖHLMANN
H. STILLER

Contents

1. Introduction

D. Möhlmann

Origin and early evolution of the planetary system are an old but not yet answered scientific problem. In spite of the fact that the basic physical theories are well known for a long time, which have to be used to describe these evolutionary processes, there was no essential progress in this field. This was caused mainly by the great uncertainties in the knowledge of appropriated initial and boundary conditions for a mathematical description of these physical processes and by the impossibility, to model numerically the complete sequence of processes from the collapse of an interstellar cloud towards the formation of stars and connected planetary systems. Only special aspects of this sequence (with uncertainties about their reality) have been investigated more in detail. Consequently, more precise astronomical data and facts on the planetary system were necessary for a further progress. These data become increasingly available with the development of space research. It is the aim of this book to include those new data and to propose a scenario, which can be applied to understand origin and early evolution of the planetary system and its evolved satellite systems of Jupiter, Saturn and Uranus as a consequence of a comparable process-sequence. Of course, the proposed planetogonic scenario needs a detailed description of the assumed process-sequence. Within this book there are only first attemps to describe some of these processes in detail. These detailed investigations have to be continued. It is therefore also the aim of this book to make the ideas of the proposed planetogonic scenario available for a greater interested scientific community and to initiate a broader discussion as well of the scenario in general as of connected detailed investigations.

1.1. Historical Background

The question of the origin of our planetary system is the central task for "planetary cosmogony" or "planetogony". Before the general acceptance of the heliocentric system of Copernicus, this question had no quantitative scientific meaning and only first qualitative or "philosophical" approaches were given within the frame of a general "cosmogony". It is remarkable that some ancient Greeks, as the Stoics and Epicurus, in spite of the lack of any realistic knowledge of the structures of the cosmos or the Solar system gave philosophical approaches, including evolutionary scenarios, which have similarities with our actual "lines of thinking". Of course, there were also the opposite points of view, neglecting any evolution, assuming with Plato and Aristoteles that everything was eternally permanent in the heavens. The first scientific basis for planetary cosmogony was given by Copernicus, Kepler and Galilei by providing us with first facts of the real structure of (parts of) the Solar system, by Newton with his formulation of the basic laws of mechanics and of gravitation and by DESCARTES (1644), who tried to understand the world as a mechanical process by assuming that laws of nature (permitting in his concept only vortex motions) would select from all (previous) chaotic configurations (of colliding particles) those that have the greatest stability (the nearly circular motions of planetary bodies). Insofar, Descartes has to be understood as the father of evolutionary planetogonic scenarios.

These ideas were continued in three directions, which can be found in more or less modified and actualized versions also in modern planetary cosmogony. They are connected with the names of BUFFON (1749), KANT (1755) and LAPLACE (1796). BUFFON (1749) assumed that the preplanetary matter was ejected from the Sun due to a collision with a comet. These "catastrophe-theories" have been modified lateron by replacing the cometary collision with a near encounter of a star. The main disadvantages of these "catastrophe-theories" are the great improbability of such encounters, the difficulties to understand the chemical and isotopic composition of planetary matter (especially the immanent dust contribution and the presence of deuterium in planets) and, last but not least, angular-momentum and energy considerations which give strong constraints to the evolution of an originally hot cigar-shaped filament of ejected solar (or stellar) matter.

The most essential break-through in planetary cosmogony came from KANT's (1755) concept of a primordial nebula as a chaotic (or nonstructured) predecessor of the Solar system. Based on Newtons laws of mechanics and gravitation and following then Descartes's idea of cosmic evolution along the well defined laws of nature, he was able to give a well founded evolutionary planetogonic scenario for the formation of the Sun and planets as a result of the interplay between (gravitational) attraction and (postulated) repulsion acting on the (cold) particles of a contracting primordial nebula. Up to now, this basic approach survived the centuries and it was the starting point for a great number of detailed and quantitative planetogonic models. A slight modification of this approach came into discussion in the first half of our century by models assuming that the evolved Sun captured cold interplanetary matter as the main source for preplanetary matter, suggesting on this way a non-cogenetic origin of Sun and planets (SCHMIDT , 1959; ALFVÉN, 1954).

A third principal approach was formulated by LAPLACE (1796), who started from a gravitationally unstable Sun instead of a chaotic primordial nebula as the source of preplanetary matter. LAPLACE (1796) assumed an originally very extended Sun with a hot atmosphere. Caused by the cooling of the atmosphere, the Sun contracted, getting due to angular momentum conservation an increasing rotational velocity until centrifugal force at the equator surpassed the gravitational force. A part of matter was left behind then from the shrinking Sun, producing discrete rings or a thin disk of hot preplanetary matter. The main disadvantages of this model, which dominated the planetogonic discussion for centuries, are essential difficulties with the real distribution of angular momentum in the Solar system and the thermal history of solid matter (dust, grains, chondrules) in the preplanetary system.

A summarizing overview on the work of the main contributors to a better understanding of planetary cosmogony, what in all cases was based on the lines of the given above basic approaches and by including methods and results of modern physics, as thermodynamics, cosmochemistry, astrophysics, plasma-physics and geophysics has been given by Brahic (1982). The interested reader is referred to that excellent overview.

As a remarkable matter of fact it must be stated that up to now there is no generally accepted consensus even on that basic approach which has to be used as the basis for further detailed work. Nevertheless, detailed investigations were made more or less in all the discussed above approaches and their modifications. This unsatisfying situation is connected mainly with the lack of precise astronomical initial and boundary conditions for planetogonic models and with difficulties to extract from the increasing amount of data on the planetary system those, which "survived" its highly irreversible formation processes and permit clear and precise information on these process-sequences. But, it should be noted, that as a first possible common tendency in modern planetogony a disk of preplanetary matter (as an intermediate preplanetary state) of dust and gas appears increasingly in newer versions of all the three approaches.

So, the main planetogonic investigations seem to have specified in two directions:

— origin and early evolution of interstellar matter towards a central body with a preplanetary disk, and
— further evolution (structurization) of preplanetary disks, bands and rings towards planetary bodies.

1.2. Astronomical Observations and Astrophysical Models

There is no doubt today that stars form in contracting parts of dense interstellar clouds, and there are also well studied special stellar types representing young stars, as bipolar nebulae, Becklin-Neugebauer and Herbig-Haro objects, T-Tauri stars and FU Orionis stars (see 5.1.6.).

Unfortunately, it is impossible up to now to observe directly earlier stellar formation processes. Consequently, initial conditions for early preplanetary processes, possibly connected with the evolution towards protostars can not be determined directly by astronomical methods, leaving yet open for discussion both types of models of co-genetic origin of stars and planets or of a non co-genetic one. This lack of information is caused mainly by the very restricted possibilities to observe interstellar matter in regions with densities higher than 10^{15} m^{-3}, where these prestellar accumulation processes proceed. Only later, at stellar densities, the matter becomes observable again in its state of newly formed protostars. This lack of exact data about the formation processes within more than 10 orders of magnitude in density is the main cause for uncertainties in astronomical initial and boundary conditions for planetogonic models.

On the other side, there are numerous serious attempts to model the gravitationally caused processes of star formation, especially by hydrodynamic codes. But the uncertainties involving viscosity, energy dissipation and the role of magnetic fields are severe. There are — especially in axi-symmetric two-dimensional models — indications for ring formation as a consequence of angular momentum conservation and for a decay of these (transient) rings to newly collapsing and ring-generating structures, offering a plausible (but not generally accepted) solution for the angular momentum and fragmentation (and mass spectrum) problems in star formation by redistributing the cloud spin to orbital angular momentum of hierarchically generated fragments (Larson, 1972; Bodenheimer, 1978; Bodenheimer and Black, 1979). However, threedimensional (simulation-) computations did not confirm these results in detail (Larson, 1977; Wood, 1981) and gave new aspects to fragmentation, as bar- and ring-instabilities.

In principle, up to now, no hydrodynamic calculation was able to follow the furthcoming evolution from a fragment after the fragmentation sequence to a rotating protostar, supposed to be in the center of a nebular disk. This disk could be the source of preplanetary matter (Kantian approach). The only existing calculations of the later stages of protostellar evolution have all assumed spherical symmetry (no rotation), but it shall be assumed that the effects of rotation, viscosity and magnetic fields may give rise only for special quantitative corrections while the qualitative results from the spherically symmetric calculations remain correct. The following qualitative results have been found from these calculations (Larson, 1978; Newman and Winkler, 1980; see 5.1.5.):

— the collapse of an appropriated fragment of interstellar clouds is non-homologous. This is because the densest region (the core) collapses fastest, while the outer part of the fragment is retarded relatively to the collapse of the core,

— the inner regions become opaque for a sufficient high density. This is followed by heating of the core since radiative energy transport becomes ineffective. At the beginning, this core formation can be of oscillatory character before a stable core forms (Tscharnuter, 1985). This hot core is surrounded at first by an extended infalling envelope that still contains most of the mass. The matter in the envelope falls into the core through an accretion shock at its surface,

— the accreting core is essentially a hydrostatic protostar. Its properties and evolution depend on those of the infalling envelope and the accretion rate of infalling matter,

— radiative energy transfer remains unimportant in the core during the entire accretion process. An outer convection zone appears during the final phases of core accretion. This might be of interest for dynamo-generated early magnetic fields.

— at later stages the infalling material may become so opaque that no significant radiative energy leaves. Then, at the end of the accretion phase, a core of a 1 $M_\odot$ protostar attains a large radius of $\sim 100\ R_\odot$ and a correspondingly high luminosity of $\sim 500\ L_\odot$. If on the other hand, the collapse is strongly non-homologous, the envelope may remain sufficiently optically thin at infrared wavelengths to allow radiation to escape from the surface of the accreting core. In this case, and if the accretion time scale is as long as the initial free fall time the core accretion ends with a radius of a few $R_\odot$ and a luminosity of a few $L_\odot$.

— the final properties of the core depend strongly on the collapse time of the envelope, what is of the order of 10^5 years and depends on both the initial density and the degree to which the collapse is non-homologous and leaves an extended envelope around the core. This can be essential also for the initial conditions of preplanetary processes and indicates a wide range for the possible properties of preplanetary circumprotostellar matter.

— the core collapses further until central temperature and density are sufficient for thermonuclear ignition. During this final collapse of the protostar the luminosity of the core ($\sim R^2T^4$) decreases. Consequently, as the temperature in the circumprotostellar shell decreases, one can expect an increase of the condensation and mineralization rates of the preplanetary matter,

— accretion can be stopped finally by radiation pressure and stellar winds when the energy, released by thermonuclear processes reaches the stellar surface about 10^6 years after thermonuclear ignition (see 5.1.6.). Then a new hydrostatic equilibrium establishes with increased surface temperature, radius and luminosity, corresponding to their HRD-values.

The above given results were derived for spherical-symmetric collapse processes, neglecting the influence of rotation. It has been shown by more actual computations (Cameron, 1985; Boss, 1985; Morfill et al., 1985) that it is especially this influence of rotation which flattens the collapsing region to a "contracting disk". In principle, the above described processes are working analogously in this disk, only the characteristical time-scales can be prolonged. In this case, the planetogonic process-sequence, as it has been described in chapter 5.3., can have been effective very early in this evolving protostellar disk even before the central body has finished its formation. It is interesting to note, that the results of the derived planetogonic scenario (see chapter 5.3.) are in favour of this early action of planetogonic processes.

Observational data which can be related to the given above scenario have been found from investigations of the environment of evolving protostars and young stars. They show for T-Tauri stars extremely active circumstellar plasma environments with strong stellar winds, accreting or expanding shells or disks, ejections or accelerations of discrete cloudlets, which seem to be related with intensive radially outward moving maser sources (with masses of planetary orders of magnitude) and complex radio structures. Furthermore they show surrounding extended circumstellar shells or disks of dust.

Following GÖTZ (1984), the characteristical radii of circumstellar plasma-shells are of the orders of 10^{10} m to 10^{11} m, they increase with their pre-main sequence age. The ratio between the radius of the (yet collapsing) protostar and its final main sequence radius ranges between 1 and 6, it decreases with the pre-main sequence age. The densities of the circumstellar shell are of the order of 10^{-11} kg m^{-3}, those of the electrons are between 10^{15} m^{-3} and 10^{16} m^{-3}. Both decrease with increasing age. The angular velocities range between 10^{-4} s^{-1} and 10^{-6} s^{-1}, also they decrease with increasing age. The magnetic induction within the shells is supposed to be of the order of (10^1-10^2) Gauss. The stellar temperature ranges between $(3-5) \cdot 10^3$ K while the temperature within the shell-plasma is of about 10^4 K. It decreases with age (GÖTZ, 1984).

A further observational result, which is related with very active phenomena in T-Tauri shells shall be mentioned here. FEIGELSON (1982) reports observations of X-ray emissions from T-Tauri and other low-mass pre-main sequence stars, indicating the presence of flares in the circumstellar shells, which are $\sim 4 \cdot 10^3$ times stronger than the largest flares of the contemporary Sun. The proton flux from such protostar-flares during the (10^6-10^7) years of pre-main sequence evolution would be sufficient to account for the ^{26}Al anomaly in meteorites (which will be discussed more in detail in part 3).

Some gross-characteristics of the structures and properties of the outer circumstellar matter surrounding young stars can be found from observations of biconical nebula. The biconical nebula S 106 (AFGL 2584) in Cygnus is one of these objects, being almost certainly a region of star formation since it is associated with a massive molecular cloud ($2.6 \cdot 10^4 M_\odot$) and H_2O- and OH masers (GEHRZ et al., 1982; see 5.1.6.). A conceptual sketch of the geometry of

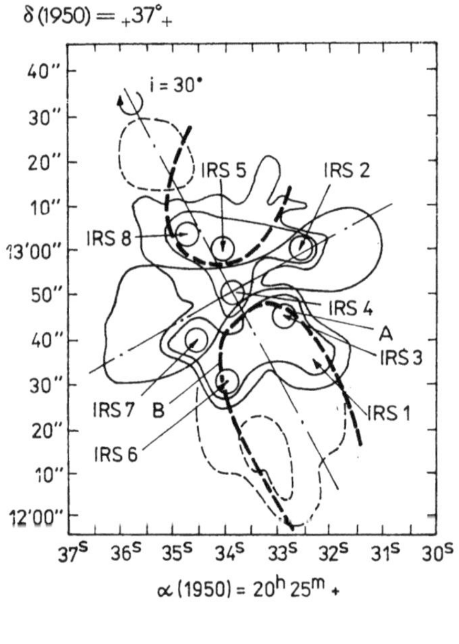

Fig. 1.1. Schematical sketch of S 106 (after GEHRZ et al., 1982).

S 106 is given in Fig. 1.1 (after GEHRZ et al., 1982). There is strong evidence from infrared and radio observations that a single 07—09 star (probabely IRS 4) is the energy source for the entire infrared, optical and radio emissions of this biconical nebula. The general morphology of S 106 suggets that this biconical nebula is inclined about 30° to NW with the north and south lobes separated by a lane of a dense irregularely shaped disk of optically opaque dust. Most of the infrared knots and the compact radio structures are associated with the heavi dashed lines which indicate shell shaped outward propagating ionization fronts, identified by ISRAEL and FELLI (1978). The infrared knots (IRS 2 and 7) seem to lie partially within the central dust disk.

Having in mind the question for planetogonically relevant information from astronomical results, the following conclusions can be drawn:

— the optical biconical and infrared appearance suggests that the exciting stars is surrounded by a dense toroidal equatorial shell (disk) which constraints the flux to escape only through the polar regions while obscuring simultaneously the central source,

— the diameter of the dust disk is $\sim 2 \cdot 10^{15}$ m, its thickness $\sim 6 \cdot 10^{14}$ m, and with a particle density of about 20 m^{-3} the total dust disk mass is $\sim 10^{-2} M_{\odot}$,

— rotation of the entire region appears to have been important in determerming the geometry of the collapsed configuration,

— non-stellar distant companions (10^4 AU or more) or not gravitationally bound objects of planetary dimensions ($10^{-2} M_{\odot} - 10^{-6} M_{\odot}$) may evolve in direct connection with star formation and might be a not yet discovered interstellar population of "first generation" planets (eventually early "brown dwarfs"), which originate directly from interstellar matter. This early process, which is not restricted to the disk-plane might be of interest also for theories of the origin of comets.

This picture of a young biconical nebula has been confirmed by observations of similar objects as Minkowski's footprint and properties of the Becklin-Neugebauer object and the Kleinman-Low nebula in the Orion nebula (OMC 2).

Detailed information concerning physical conditions (as temperature, pressure and particle sizes) and chemical state and composition of circumstellar disks as possible predecessors of planetary systems are not precisely known from astronomical investigations. Future astronomical work is necessary to answer these questions exactly. On the other side, during the last twenty years cosmochemical investigations of the matter in our planetary system gave essential insight in the physico-chemical state of the preplanetary disk. These results will be discussed in the next section, and more detail in part 3.

Summarizing the above mentioned results of astronomical observations and astrophysical models it can be stated that a circumstellar disk is a natural by-product of the evolution towards protostars and of protostars. Depending on the initial conditions, determined by the yet unknown details of the fragmentation process, there is a wide range of parameters, as temperature, luminosity and density for the protostar and its environment. Physical processes in the circumstellar shell can be characterized by

- collapse-caused density increase and cooling of the outer circumstellar matter for some (10^5—10^6) years until the energy released by nuclear reactions leaves the stellar surface, what could be connected with condensation- and mineralization processes, giving a second type of solid preplanetary matter after the earlier solids of interstellar or stellar origin.
- highly violent processes due to stellar winds, giant stellar flares and radiation-heating from the young star. These processes could be a cause for reheatings of the solids, local nuclear reactions due to high energetic protons and an outward streaming of the gaseous component of circumstellar matter.

Consequently, and as will be discussed more in detail in part 3, the physical, chemical and mineralogic state of preplanetary matter must assumed to be very complex. But, in spite of all complexity, the astronomical observations show further that a thin preplanetary circumprotostellar disk can be assumed as an initial phase for more detailed models of the (possibly very early) evolution towards planetary bodies (protoplanets). This will be discussed more in detail in part 5.

1.3. Meteorites and Space Research Results

A better understanding of the history of the Solar system became possible during the last twenty years as well by direct investigations in the planetary system as by improved analysis of meteorites. It should be mentioned, that these improvements in the capability for high precision measurements of meteorites were made possible by significant technological advances of space-research and in connection with the study of lunar rocks. It is the main task of these matter-related investigations to understand the evolutionary steps of planetary matter from their beginning in circumstellar shells (for details see part 3).

In principle, and having in mind the above given results of models and observations of protostars and their circumstellar shells, the following sources for solid preplanetary matter should have been effective:

- condensation and solidification of hot matter to dust in circumstellar shells of late stars, novae and supernovae, providing the interstellar clouds with dust,
- condensation and solidification of cold interstellar gases in connection with preexisting interstellar dust grains as nucleation sites,
- condensation and solidification of hot matter (plasma), ejected by the protostar,
- condensation and solidification of cool matter in course of the protostellar collapse in connection with cooling and collapse-caused increase in density in outer parts of circumprotostellar matter.

This scenario has to be modified by reheating processes, possible evaporation (and repeated condensation) due to heating from the collapse of the protostar and finally from the evolved star and its giant flares. These processes produce further modifications of the primitive microsized solid matter so that the original sources can not be identified directly in every case. A further complication arises from the petrologic evolution from dust to macroscopic bodies where further irreversible differentiation took place.

Nevertheless important information on preplanetary matter has been derived from detailed investigations of the most primitive meteorites, the chondritic meteorites. They are stony meteorites, containing millimeter-sized silicatic chondrules, appearing like solidificated droplets of a melt and being imbedded in a dark matrix, which includes also metallic and FeS particles. The matrix is more oxidized than the chondrules. It contains an important amount of low-temperature volatile phases and is very similar in composition to the solar corona, indicating that it can be a cold agglomeration of matter in outer regions of the early circumprotosolar disk. The chemical composition could be that of the primordial fragment of an earlier interstellar cloud.

Rb—Sr datings show that the chondritic meteorites are (4.55 ± 0.015) billion years old. Eventually, the uncertainty of this formation time can be reduced. This could indicate a yet shorter time interval for these petrologic processes of the growth of primitive meteorites from a differentiated earlier state of dust, grains and gas-plasma. The fact, that no significant heating modified the chondritic meteorites since their origin, makes it plausible, that they are relicts of these earlier condensation — and minerogenetic and petrologic processes and that their chemistry and mineralogic state were established during the early phases of the genesis of solid preplanetary matter in the circumprotostellar disk. It should be mentioned furthermore, that in spite of their short formation time, these primitive meteorits should have been grown "gently" by "sinking together" instead of violent collisions. This can be an indication for a very early growth of these bodies in a dust-gas environment (and not by collisional accretion).

The following detailed results have been derived from laboratory investigations of primitive meteorites:

- A very remarkable result of these investigations in combination with astronomical observations was to confirm that the deuterium abundance, or more precise, the deuterium/hydrogen ratio is greater by two orders of magnitude or more in interstellar matter, meteorites and major planets than in the solar photosphere. This can be understood by the fact, that the nucleus of deuterium is barely bound (proton and neutron are almost outside the range of the nuclear forces). Consequently, deuterium will be destroyed in short timescales at temperature of 10^6 K or greater. This implies that not all of the preplanetary matter can have been ejected from its stellar central body, eleminating therefore the above mentioned corresponding models of a pure solar origin from further discussion,
- there are appearently noble gas components in meteorites of low petrologic type that experienced little or not metamorphism. It is curious that the

presolar (interstellar) phases are of larger grain size than the "solar" phases, suggesting low nucleation rates and long growth times (ANDERS, 1981). Here it should be noted, that the existence (survival) of presolar grains gives strong evidence as well for relatively low temperature of the outer parts of the preplanetary nebula (excluding also a pure solar origin) as for a restricted influence of later stellar winds on (at least parts) the solid preplanetary matter. An alternative approach could be only the later capture of preplanetary matter by the evolved Sun (ALFVÉN, ARRHENIUS, 1976),

- micron-sized grains ("nuggets") in Allende contain all refractory (Mo, Re, W) and Pt-group (Pt, Os, Ir, Ru, Rb) metals in solar proportions relative to each other and highly enriched to the other elements (LATTIMER, 1982). They are obviously simple unfractionated initial condensates which have not melted or reacted chemically. They are probabely condensed from ejections of the protosun,
- magnetic grains in carbonaceous chondrites show a remanent magnetization. The magnetic alignment of the aggregates indicates that growth and/or aggregation took place in magnetic fields of the order (10^{-5}—10^{-4}) Vs/m^2. This observed magnetization in primitive meteorites can not derive planetary fields (BRECHER, 1971; BRECHER and ARRHENIUS, 1974), but it could be related with a large-scale magnetic field from the evolving central body,
- isotopic anomalies in meteorites seem to indicate that there is presolar information especially in primitive meteorites:

  a) The ^{22}Ne excesses in the matrix material (of the Orgueil meteorite) led to the conclusion of a presolar nucleosynthetic origin (BLACK, 1972), but having in mind the results of FEIGELSON (1982), suggesting strong high energetic proton fluxes from giant flares of pre-main-sequence stars, the origin of ^{22}Ne in early preplanetary phases can not be excluded.

  b) The ^{16}O excesses, found in inclusions of the Allende meteorite (CLAYTON et al., 1976) indicate, that the oxygen in some minerals of these meteorites is not in isotopic equilibrium with the surrounding meteoritic matter. As there are no parent radioactive isotopes ^{16}O, this oxygen isotope can have been produced at nucleosynthese sites only. Lateron it should have been added to the protosolar nebula by presolar grains. But, as a remarkable matter of fact, there is (with two exceptions only) a lack of correlation of ^{16}O anomalies with other anomalies. This might be in correlation with the results of THIEMER and HEIDENREICH (1983), indicating an influence of intense UV-radiation on the chemical fractionation of oxygen isotopes. Up to now, this is not finally understood.

  c) Similar conclusions about a protosolar origin can be found from investigations of the ^{26}Mg excesses in some inclusions. They are assumed to be caused by an in situ decay of ^{26}Al (half life 750000 years), indicating either that these ^{26}Mg rich grains condensed outside the proto-

solar nebula at the site of the ^{26}Al production (i.e. near to a supernova), or that freshly synthesized ^{26}Al has been incorporated to the preplanetary matter, where high temperature condensates are enriched then by ^{26}Al. As a further possibility the ^{26}Al production by irradiation (Heyman, Dziczkaniec, 1976, Feigelson, 1982) has to be taken into account more seriously.
A further approach to understand the ^{26}Mg anomalies considers the ^{26}Al as homogeneously distributed along with ^{27}Al over the solar nebula gas, out of which the refractoring grains condensed (Bibring, Langevin, 1982). This implies a high ^{26}Al content in the preplanetary matter, being sufficient to melt the interior of planetary bodies down to 10 km in diameter (Lee et al., 1976) due to the heat, released by the rapid decay. This might have been of great planetogonic relevance.

Obviously, the interpretation of the information from isotopic investigations is very complex and has not finished. But two aspects of the resulting conclusions shall be repeated here:

- very probably presolar grains were a constituent of preplanetary matter in the protostellar nebula,
- isotope anomalies indicate that the presolar nebula was not totally homogeneous, it can not have been heated and mixed completely (excluding therefore the corresponding homogeneous planetogonic models).

Summarizing the results of matter related investigations (which will be discussed more in detail in part 3), it must be stated, that origin and growth of solid preplanetary matter can not be described within a picture of a single one-way road. Obviously there was a number of different coupled and crossing ways of evolution, probabely starting from the connected main road of stellar and interstellar gas and condensations and further condensation, solidification and reheating (Kurat, 1970) in the cooling dense circumstellar environment, also from hot matter of ejections from the protostar, and possibly connected nuclear reactions. Consequently, the Kantian approach to planetogony seems to be most appropriated. But it has to be modified as well by taking into account contributions to preplanetary matter, ejected from the protostar, as by the action of early magnetic fields of protostellar origin.

Another remarkable result which follows from matter-related planetogonic investigations is that the solid matter in the Solar system must have been formed under strongly reducing conditions (as they are given in a hydrogen-dominated circumprotostellar preplanetary gas and in protoplanets). This may be an independent argument for the formation of solid matter and planets in a gaseous disk and via the phase of protoplanets.

A further result is, that planetary bodies and resulting global differentiation established as early as $4.55 \cdot 10^9$ yrs ago (Allegre, 1982; Tera and Wasserburg, 1974; Crozaz et al. 1972). Furthermore, the common ages of Earth, Moon and meteorites are an indication for a relatively fast process of the formation of planetary sized bodies (at least in the inner Solar system). This seems to be in agreement with ages of $4.52 \cdot 10^9$ yrs of some highly differentiated meteorites

(LUCK et al., 1980), which were separated by catastrophic events from highly differentiated evolved parent bodies with a core (iron meteorites) and surface volcanism (achondrites). Here it should be noted, that this fast growth seems not to be compatible with models of a stochastic (and consequently slow) growth from statistically colliding and hierarchically growing planetesimals.

This necessity of a faster planetogonic process-sequence was one of the motivations to discuss planetogony anew by the authors. Therefore, and based on the above given facts and interpretations, a planetogonic scenario with a fast growth of planetary bodies from gas, dust and grains, focussed by collective effects, as instabilities of the disk, has been developed more in detail and in competition with processes of a stochastic growth. This scenario of a probabely very early collective fast evolution towards planetary bodies is the central idea of the planetogonic approach described in this book. A more detailed description of relevant physical processes will be given in part 5.

1.4. Reconstruction Characteristica and Relevant Processes

As has been discussed above, astronomy and cosmochemistry are able to give increasingly more detailed data needed to attack one of the two above formulated tasks for planetogony, namely to understand "origin and early evolution of interstellar matter towards a central body with a preplanetary disk". This has been discussed in the foregoing section and is partially the subject of parts 3 and 5. The other task, to study "further evolution (structurization) of circumprotostellar preplanetary disks, bands and rings towards planetary bodies" needs the knowledge of structures in the Solar system. This is the task for "comparative planetology". Here it seems to be of great importance that the Solar system has four developed systems of a central body and satellite system, namely the planetary system and the satellite systems of Jupiter, Saturn and Uranus. The postulate, that common and comparable properties and structures of these systems are of planetogonic relevance, can be used then as a criterion to identify characteristic parameters of the planetogonic formation processes and to separate them from peculiar or "individual" properties. Vice versa, the verification of the existence of such common and comparable characteristics is a strong argument in favour of this assumption of a comparable evolution of preplanetary and presatellite disks. Characteristics of the above described type will be used as "reconstruction characteristica". They combine the ideas of the "hetegony principle" of ALFVÉN and ARRHENIUS (1976) with the assumption of a "common formation principle" (STILLER, TREDER, MÖHLMANN, 1981). It is interesting to note, that the existence of comparable gross-structures in the four systems indicates that the processes of their formation were independent of special initial- and boundary conditions (having asymptotically stationary final states). Well known in literature are the "classical" reconstruction characteristics

- coplanarity of the orbits,
- circularity of the orbits,
- anisotropy due to the dominance of one angular-momenta orientation in every system,

which are strong indications for a preplanetary thin disk and for a common origin (separation) of the matter of every system from a greater structure, as it is implied by the dominance of one angular-momentum orientation (in spin and orbits) for every system.

A further reconstruction characteristicum is the "quasiperiodic" distribution of the orbital radii in the four systems. This quasiperiodicity is a key for the mathematical modelling of planetogonic processes in part 5. Therefore, these radial structures shall be discussed more in detail.

Table 1.1. Radial structures in the planetary system

Planet	Semimajor axis (10^{11} m)	Orbital velocity (km/s)	Radial distance (10^{11} m)	Velocity diff. (km/s)	Orbital radius (10^{11} m)* acc. structure model
Mercury	0.58	47.9			0.62
?	1.08		0.5	2 · 6.4	0.80
Venus	1.08	35.1	0.42	5.3	1.07
Earth	1.50	29.8	0.78	5.7	1.51
Mars	2.28	24.1	1.42	5.2	2.27
Asteroids	3.7	18.9	4.08	5.8	3.80
Jupiter	7.78	13.1	6.52	3.5	7.62
Saturn	14.3	9.6	14.4	2.8	14.2
Uranus	28.7	6.8	16.3	1.4	29.7
Neptune	45.0	5.4	14.5	0.7	44.4
Pluto	59.5	4.7			59.5

* For the inner system (from Mercury to Jupiter) the orbital radii have been determined by assuming a structure with constant velocity-differences ($\Delta v = 5.5$ km/s) and an "original" orbital velocity of the Earth from the average of that of the actual values for Venus, Earth and Mars ($\bar{v}_0^{(E)} = 29.7$ km/s). The orbital radii follow then from $r = \gamma M_c/v_0^2$. The orbital radii of the outer system have been computed by assuming a constant radial distance between the planets ($\bar{\Delta} r = 1.51 \cdot 10^{12}$ m) and an "original" orbital radius of Uranus, determined from the average of that of the actual orbital radii of Saturn, Uranus and Neptune ($r_u = 2.93 \cdot 10^{12}$ m).

The radial structures in the planetary system are described with Table 1.1. It can be seen that the "inner planetary system" (between Mercury and Jupiter) can be characterized by a constant velocity-difference ($\Delta v = 5.5$ km/s). It shall be noted here, that this "spacing" by constant velocity-differences has been discussed first by LITZROTH (1980) followed by MÖHLMANN (1981). In the same way it can be found from Table 1.1. that the "outer planetary system" (from Saturn to Pluto) can be characterized by a constant radial distance of about $15.1 \cdot 10^{11}$ m between the planetary orbital radii (MÖHLMANN, 1985). These results indicate that the planetary system can be separated

Tabelle 1.2. Radial structures in the Jovian system

Satellite	Semimajor axis (10^8 m)	Orbital velocity (km/s)	Radial dist. (10^8 m)	Velocity diff. (km/s)	Orbital radius (10^8 m) (acc. structure model*
Callisto	18.8	8.21	8.1	2.69	19.4
Ganymede	10.7	10.9			10.1
			3.99	2.80	
Europa	6.71	13.7			6.29
			2.49	3.60	
Io	4.22	17.3			4.23
			· 2,0	2 · 3.3	3.05
J2	2.22	23.9			2.31
			0.41	2.6	
Amalthea	1.81	26.5			1.80
			0.47	4.3	
J1	1.23	32.1			1.45
			0.16	2.0	

* It should be noted, that only satellites with low inclinations were taken into account.
The orbital radii have been determined by assuming a structure with constant velocity-differences ($\bar{\Delta}_v$ = 3.07 km/s) and with the velocity of Amalthea (v_A = 26.5 km/s) as a references value. The Galilean satellites have not been used to find a reference level. This is due to the resonances between these satellites.

into an inner system and an outer system with different radial structures. This can be an indication for different preplanetary conditions in these two parts of the preplanetary disk.

The realitiy of orbital spacing by constant velocity-differences is strengthened by the structures in the satellite systems of Jupiter, Saturn and Uranus, as they are described by the Tables 1.2, 1.3 and 1.4. Indeed, all these systems can be described on this way of "velocity-determined" spacing.

It is interesting to note, that on the basis of this structuremodel for planetary radii, predictions were possible for the existence of orbital level for Uranian satellites (Möhlmann, 1984c) which were found later in 1986 by the Voyager-spacecrafts.

Further reconstruction characteristica seem to be

— the position of the inner boundaries of the systems
— the radial distribution of mass and composition of the satellites and planets (which indicates too, that the Solar system separated into an inner and an outer system)
— spin-periods of the free rotating bodies (spin-isochronism).

Table 1.3. Radial structure in the Saturnian system

Satellite	Semimajor axis (10^8 m)	Orbital velocity (km/s)	Radial dist. (10^8 m)	Velocity diff. (km/s)	Orbital radius (10^8 m) acc. structure model*
Japetus	35.61	3.26			27.2
			20.8	1.8	
Hyperion	14.81	5.06			
			2.59	0.51	13.8
Titan	12.22	5.57			
			6.95	2 · 1.46	8.27
Rhea	5.27	8.48			5.52
			1.50	1.52	
Dione	3.77	10.0			3.94
			0.82	1.3	
Tethys	2.95	11.3			2.95
			1.3	1.3	
Enceladus	2.38	12.6			2.30
			0.57	1.7	
Mimas	1.86	14.3			1.84
			0.35	1.5	
S1, S3	1.51	15.8			
			0.11	0.7	1.50
S26, S27, S28	1.40	16.5			

* The orbital radii have been determined by assuming a structure with constant velocity – differences ($\Delta \bar{v} = 1.52$ km/s) and with an "original" velocity level of of 11.33 km/s for Tethys (determined from the averages of the orbital velocities of the satellites from Mimas to Rhea). The deviations outward of Titan could be caused by the gravitational influence of this massive satellite. The predicted level at $8.27 \cdot 10^8$ m corresponds to the ring, discovered by VASUNDHARA (1984). The prediction has been mode by MÖHLMANN (1984c).

As will be discussed more in detail, in part 5, the position of the inner boundary can be understood in connection with the evolution towards a thin disk, influenced by the action of gravitational, centrifugal and magnetic forces, resulting in a partial corotation of the early gas-plasma (ALFVÉN, ARRHENIUS, 1976). As a yet open problem in this picture remains the generation of a strong large-scale magnetic field in a contracting and rotating protostar (before thermonuclear ignition) and its predecessing disk, or in a rapidly forming protoplanet. This seems to be a planetogonically highly relevant task for the dynamo-theory of planetary and stellar magnetic fields. Convection zones, as they are necessary for dynamo action, seem to evolve during proto-

Table 1.4. Radial structures in the Uranian system

Satellite	Semimajor axis (10^8 m)		Orbital velocity (km/s)	Radial dist. (10^8 m)	Velocity diff. (km/s)	Orbital radius (10^8 m) acc. structure model*
Oberon	5.83		3.15			8.35
				1.48	0.5	
Titania	4.36		3.65			4.38
				1.71	1.02	
Umbriel	2.66		4.67			2.68
				0.75	0.84	
Ariel	1.91		5.51			1.80
				0.62	1.18	
Miranda	1.30		6.69			1.30
?						0.98
1986 U4;	0.6992;					
1986 U5;	0.753;				4×0.998	
1985 U1	0.860	(0.77)				0.77
1986 U9;	0,592;					
1986 U3;	0.618					
1986 U6;	0.627;					
1986 U2;	0.646;					
1986 U1	0.661	(0.63)				0.62
1986 U8	0.538					
ε-ring	0.512	(0.51)	10.7			0.51
1986 U7	0.497					

* The orbital radii have been determined by assuming a structure with constant velocity — differences ($\Delta\bar{v} = 1.01$ km/s) and with the orbital velocity of Titania as the reference value. The averages of the orbital radii of the newly found Uranian satellites are given in brackets for the three groups respectively. It is interesting to note, that these radii agree with those from the structure model. These levels for orbital radii in the inner parts of the Uranian system have been predicted by MÖHLMANN (1984c, 1985) before the discovery of these satellites by the Voyager — spececrafts in 1986.

stellar evolution (see part 5), but for protoplanetary conditions this problem is yet open too.

Having now the model picture of a preplanetary disk of gas, dust and grains as an initial state and the radial quasiperiodicity of orbital radii as a final state, which is comparable in all the four above mentioned systems, the question for a fast planetogonic process leads to the investigation of the radial gravitational stability of disks in the gravitational field of a (growing) central body. This investigation has been made in part 5, giving clear indications for the (fast) instability of these disks even in the frame of linear perturbation theory. First hints on this instability even under restricted conditions can be found in the literature (TOOMRE, 1964; HUNTER, 1965; POLYATCHENKO and FRIDMAN, 1972). The resulting decay of the disk leads to radially quasiperiodically spaced rings or bands, where the radial spacing depends on ratio of central mass and preplanetary mass distribution (disk-surface-density) and on the thickness of the disk. Insofar, it could to be possible to get some information about the early surface density from the actual spacing of the planetary- and satellite orbits, if it can be assumed that planets formed at the sites of the rings or bands with locally preenhanced density (in comparison with that of the disk). This assumption seems to be justified since rings can shown to be unstable, indicating the possibility for a rapid gravitational collective accretion to one or a few "protoplanetary" bodies in this region. It should be noted here, that this idea of an early protoplanetary state in preplanetary evolution, which results from gravitational instabilities in the preplanetary disk, has been advocated by CAMERON (1973) even for a long time. The "azimuthal instability", causing this process of rapid collective accretion due to self-gravitation in rings is discussed more in detail in part 5. First indications for the existence of such an instability can be found in the literature (Boss, 1982).

This model of preplanetary evolution can be confirmed by the fact, that gravitationally focussed accretion from rings or bands is able to reproduce the observed property of spin-isochronism. This will be discussed in part 5.

Further information on these early processes of planetary growth and internal evolution can be found from investigating models of individual planetary evolution and structuring (i.e. towards structures as core or mantle) and by studying the energetics of different structuring mechanisms. These investigations give further contraints for planetogonic "modeling". They are discussed in detail in part 4, and there, it is especially the thermal evolution of planetary bodies of terrestrial type and the resulting possibilities for convection and magnetic fields which give indications that are in favour of a protoplanetary origin of the planets instead a growth by stochastic accretion.

The above outlined consistent planetogonic scenario shall be summarized here with its essential steps:

- accumulation of interstellar matter of gas and dust,
- fragmentation of a contracting interstellar cloud down to the level of a forming protostar with a massive circumprotostellar disk of condensed and partially reheated preplanetary matter and matter resulting form ejections from the protosun,
- possible formation of an inner boundary and magnetic focussing towards an "inner" disk in its early gas-plasma state of partial corotation (with more or less comoving presolar grains) due to a large scale stellar magnetic field,
- partial outgassing due to heating, radial outflow at greater vertical heights of this "inner" disk which yet "feeds" the protostar too,
- collective gravitational instabilities separate this early disk into well ordered quasiperiodically spaced rings or bands,
- decay of these rings or bands by the "azimuthal instability", (which becomes effective due to ring formation and related increase in density)

and direct rapid gravitational collective formation of planetary sized bodies (protoplanets).

The advantages of this scenario seem to be:

- the compatibility with the discussed above cosmochemical, astronomical and space research related results,
- the analogue applicability of this scenario on the four evolved systems of central body with satellite system,
- the possibility for predictions (in 1985) of the existence of radial ranges for the newly found Uranian satellites and possible new aspects on origin of the early Moon.

It is the aim of this book to describe more in detail some of the steps of the above outlined scenario.

2. Mechanical Properties of the Present Solar System

F. Spahn, H. Sponholz

Presently known orbital and further mechanical data of the planets and their satellites are given in this part. Tables 2.1—2.4 show them for the four systems — the planetary system and the three satellite systems of Jupiter, Saturn and Uranus, respectively. Table 2.5 gives the corresponding values for the remaining planetary sized bodies of the Solar system, namely Moon, Phobos, Deimos, Triton, Nereide and Charon.

Derived mechanical data can be obtained from these values.

References of Chapter 2

[1] Landoldt-Börnstein (1981)
[2] Belton et al. (1980)
[3] Stone and Miner (1982)
[4] Smith et al. (1982)
[5] Smith et al. (1986)
[6] IAU Circ. Nr. 4168 (1986)
[7] IAU Circ. Nr. 4164 (1986)
[8] Szeto (1983)
[9] Morrison and Samz (1980)
[10] Esposito et al. (1983)
[11] Stone and Miner (1986)

For the complete references see reference list at the end of the book.

Table 2.1. Planetary System (data from [1])

planet	semi axis a (10^8 km)	eccentricity e	inclination i	revolution period P (days; years)	mass m (10^{24} kg)	mean density (10^3 kg/m^3)	rotational period T (hours)	diameter (equatorial) (polar) D (km)	equatorial inclination * i_e
Mercury	0.579	0.2056305	7°0′15″	87.969 d	0.3302	5.43	1407.6	4878	2°
Venus	1.082	0.0067825	3°23′40″	224.701 d	4.869	5.24	5832	12104	183°
Earth	1.496	0.0167175	0.0	365.256 d	5.974	5.515	23 h 56′4″	12756.28 12713.51	23°27′
Mars	2.279	0.093387	1°50′59″	686.98 d	0.6419	3.93	24 h 37′23″	6794.4 6754.6	23°59′
Jupiter	7.784	0.048027	1°18′19″	11.869 a	1898.8	1.33	9 h 55′	142796 133800	3°4′
Saturn	14.292	0.0537	2°29′11″	29.628 a	568.4	0.70	10 h 25′	120000 106900	26°44′
Uranus	28.781	0.048	0°46′21″	84.665 a	86.98	1.27	24.4 h [2]	50800 (49780)	98°
Neptune	45.08	0.008	1°46′17″	165.49 a	102.8	1.71	15.4 h [2]	48600 (47470)	29°
Pluto	59.28	0.252	17°8′20″	251.86 a	0.015	(0.7)	(153 h)	(3500)	> 50°

* denotes the inclinations with respect to the planetary orbit!

Table 2.2. Jovian System (data from [1])

Satellite	semi axis a (10^3 km)	eccentricity e	inclination i (° ′ ″)	revolution period P (days)	diameter D (km)	mass m ($10^{22} \cdot$ kg)	density (10^3 kg/m^3)
1979 J1	127.8	0.	0°	0.294	< 40	—	—
J5 Amalthea	181.3	0.0028	0°24′	0.498	240	—	—
1979 J2	221.7	—	1°15′	0.675	≈ 80	—	—
Io	421.6	0.00…	0.00…	1.769	3650	8.92	3.5
Europa	670.9	0.0003	0°1′12″	3.551	3120	4.86	3.1
Ganymede	1070	0.0015	0°5′24″	7.155	5280	14.89	1.9
Callisto	1880	0.0075	0°25′48″	16.689	4840	10.63	1.8
Leda	11094	0.148	27°*	239	10	—	—
Himalia	11470	0.158	28°*	250.6	170	—	—
Lysithea	11710	0.130	29°*	260	24	—	—
Elara	11740	0.207	28°*	260.1	80	—	—
Ananke	20700	0.170	147°*	617	20	—	—
Carme	22350	0.210	163°*	692	30	—	—
Pasiphae	23300	0.380	148°*	735	36	—	—
Sinope	23700	0.280	153°*	758	28	—	—

* denotes the inclination with respect to the planetary orbit!
The other inclinations are taken with respect to the planetary equator.

Table 2.3. Satellites of Saturn (references are given in the table-head)

Satellite	semi axis a [3] (10³ km)	eccentricity e [1]	inclination i [1] (° ′ ″)	revolution period P [3] (days)	diameter D [3] (km)	mass m [4] (10²¹ · kg)	mean density [3] (10³ kg/m³)
1980 S28	137.67	—	0	0.602	40 × 20	—	—
1980 S27	139.53	—	0	0.613	140 × 100 × 80	—	—
1980 S26	141.70	—	0	0.6285	110 × 90 × 70	—	—
1980 S3	151.422		0	0.6943	140 × 120 × 100	—	—
1980 S1	151.472	0	0	0.6947	220 × 200 × 160	—	—
Mimas	185.54	0.0201	1°30′	0.94242	392	0.0375	1.44
Enceladus	238.04	0.0044	0	1.3702	500	0.084	1.16
Thetys	294.67	0.0...	1°6′	1.8878	1060	0.755	1 21
1980 S25	294.67	—	—	1.8878	34 × 22 × 22	—	—
1980 S13	294.67	—	—	1.8878	34 × 28 × 26	—	—
Dione	377.42	0.0022	0	2.7369	1120	1.05	1.43
1980 S6	378.06	—	0	2.7391	36 × 32 × 30	—	—
Rhea	527.10	0.0010	0°24′	4.5175	1530	2.49	1.33
Titan	1221.86	0.0289	0°18′	15.9454	5150	134.57	1.88
Hyperion	1481.0	0.1042	0°24′	21.2766	410 × 260 × 220	—	—
Iapetus	3560.8	0.0283	18°24′*	79.3308	1460	1.88	1.16
Phoebe	12954	0.1591	174°48′*	550.45	220	—	—

* inclinations with respect to the planetary orbit; the other ones are with respect to the planetary equator.

Table 2.4. Satellites of Uranus

Satellite	semi axis a [5] (10^3 km)	eccentricity e [1]	inclination i [1] (° ′ ″)	siderial period p (days)	diameter D (km)	mass m* (kg · 10^{21})	mean density [5] (10^3 kg/m^3)
1986 U7	49.7	—	—	0.33 [6]	≈ 20 [6]	—	—
1986 U8	53.8	—	—	0.3717 [6]	≈ 25 [6]	—	—
1986 U9	59.2	—	—	0.4325 [6]	≈ 50 [6]	—	—
1986 U3	61.8	—	—	0.46375 [7]	≈ 80 [7]	—	—
1986 U6	62.7	—	—	0.47458 [7]	≈ 60 [7]	—	—
1986 U2	64.6	—	—	0.492917 [7]	≈ 80 [7]	—	—
1986 U1	66.1	—	—	0.51333 [7]	≈ 100 [7]	—	—
1986 U4	69.9	—	—	0.55875 [7]	≈ 60 [7]	—	—
1986 U5	75.3	—	—	0.622083 [7]	≈ 60 [7]	—	—
1985 U1	86.0	—	—	0.761833 [7]	≈ 150 [7]	—	—
Miranda	129.783	0.017	3°24′	1.413 [1]	484 [5]	0.075	1.26
Ariel	191.239	0.0028	0	2.520 [1]	1160 [5]	1.35	1.65
Umbriel	265.696	0.0035	0	4.144 [1]	1190 [5]	1.3	1.44
Titania	435.844	0.0024	0	8.706 [1]	1610 [5]	3.5	1.59
Oberon	582.596	0.0007	0	13.46 [1]	1550 [5]	3	1.50

* The masses are calculated from the diameter and the mean density for an assumed spheric shape of the satellite.

Table 2.5. Remaining satellites of the Planetary System

Satellite	semi axis a [1] (10^3 km)	eccentricity e [1]	inclination i [1] (° ′ ″)	revolution period P [1] (days)	diameter D [1] (km)	mass m (kg)	density (10^3 kg/m^3)
Moon	384.40	0.0549	18°18′—28°36′	27.321661	3476	$7.35 \cdot 10^{22}$ [1]	3.34 [1]
Phobos	9.380	0.0150	1°6′	0.3189	27 × 22 × 19	$9.9 \cdot 10^{15}$ [8]	2.0 [8]
Deimos	23.46	0.00052	0°54′—2°42′	1.2620	15 × 12 × 11	$2.0 \cdot 10^{15}$ [8]	1.9 [8]
Triton	354.0	0.00	160	5.877	4400	$2.0 \cdot 10^{23}$ [1]	(4.6) [1]
Nereid	5510	0.75	28*	360.1	(300)	—	—
Charon	20.0	0.0	120*	6.387	1200...2000	—	—

* inclinations with respect to the planetary orbit!

Table 2.6. Jovian Ring (data from [9])
The distance is measured from the center of Jupiter

	outer edge distance [10^3 km]	width [km]	thickness [km]
ring	128.3	≈ 6500	< 30

Table 2.7. Saturnian Rings
Distances are measured from the center of Saturn (from [3])

feature	inner edge-distance [10^3 km]	outer edge-distance [10^3 km]	mean optical depth [10]
D-ring	67.0	74.4	—
C-ring	74.4	91.9	0.08—0.15
B-ring	91.9	117.4	1.2—1.9
Cassini division	≈ 117.4	≈ 121.9	0.12
A-ring	121.9	136.6	0.57—0.7
F-ring	140.3*	—	—
G-ring	170.0*	—	—
E-ring	180.0	480.0	—

* distance to the center of the ring

Table 2.8. Uranian Rings (data from [11])
Distances with respect to the center of the planet and the rings

feature	distance (10^3 km)	eccentricity e (10^{-3})	inclination (to the equator) i (10^{-3} degrees)	width (km)	mean optical depth
1986 U2R	37...39.5	(0)	(0)	≈ 2500	$10^{-3}...10^{-4}$
6-ring	41.85	1.0	63	1—3	0.2—0.3
5-ring	42.24	1.9	52	2—3	0.5—0.6
4-ring	42.58	1.1	32	2—3	0.3
α-ring	44.73	0.8	14	7—12	0.3—0.4
β-ring	45.67	0.4	5	7—12	0.2
η-ring	47.18	(0)	(2)	0—2	0.1—0.4
γ-ring	47.63	(0)	(11)	1—4	1.3—2.3
δ-ring	48.31	(0)	4	3—9	0.3—0.4
1986 U 1R	50.04	(0)	(0)	1—2	0.1
ε-ring	51.16	7.9	(1)	22—93	0.5—2.1

3. The Nature of Preplanetary Matter

R. Wäsch

During the last ten years it has been shown mainly by investigations of chondritic meteorites, that the origin of as well solid (silicates, oxides, metals) planetary material as well as planetary bodies took place in the course of rather variable thermodynamic and kinetic conditions and in different „synthetic“ sites. Especially the following conclusions have been drawn from the investigation of the distribution of chemical elements and isotopes in different meteoritic types and meteoritic constituents, by age relations, from the coexistence of meteoritic constituents having different genetic starting-points and histories (in an individual type) and by the possible existence of "differentiated" meteoritic precursors and different consolidation states (more powdery-like agglomerated or more rocky-like consolidated) of various meteorite types:

- The isotopic patterns, above all of the carbonaceous chondrites, indicate an origin of the planetary bodies of the solar system from an imperfectly mixed reservoir of matter (see section 3.4.4.).
- The formation of as well solid preplanetary material as well as planetary parent bodies of different size and evolutionary states occured almost simultaneously in a very short time interval. By radioactive dating it has become evident that most of the known meteorites are of nearly the same age (see section 3.4.1.).
- The solid precursors of planetary bodies consisted of matter which originated under quite different genetic conditions. This material was composed of dust matter formed by various processes as supernova(e) ejecta, "indigenous" solar condensates, and condensates of revaporized solid presolar material (section 3.2.).
- The agglomeration of aggregates and grains forming meteoritic or later planetary parent bodies (planetesimals of a "first generation") apparently was a gentle process. In this way, the "peaceful" neighbourhood of only weakly interacting hightemperature grains (f. e. CAI's, chondrules, etc.) and deeptemperature matrix silicates observed in some carbonaceous chondrites is explainable (section 3.4.6.).
  Additionally, the first results of the investigation of the abundance pattern of the major and some minor elements in cometary dust seem to confirm that the matrix compounds of CI-chondrites represent the most primitive preplanetary material as a matter of fact (section 3.5.).

Some experimental and theoretical results and connected interpretations which base the above drawn conclusions are based on shall be discussed and commented in the following sections of this paper.

3.1. Introduction

It is obviously not possible to deduce in a unique way the origin of the planetary system from that state of matter in which we know it today, even if we would know all about its present day state. The reason for this is that it is very difficult to identify the primary and "most primitive" preplanetary material. Up to now it is not yet clear, whether there is still such primitive matter in the solar system.

All of the terrestrial planetary bodies consist of "evolved" material which has undergone further differentiation. This seems to be the case for a great part of meteorites and, quite probably, for many asteroids too. Regarding meteorites, there is a general consensus, that the carbonaceous chondrites constitute the most primitive solar system matter available on Earth. Their constituents condensed about 4, 6 b. y. ago and have not been altered essentially since their aggregation. The chemical composition of the carbonaceous chondrites, above all the C1-types, are very similar to the elemental spectrum of the solar corona. They can be classified into several types, according to the degree of primitiveness (Table 3.1).

The C1-chondrites essentially have solar composition except for noble gases and volatiles such as H, C, N, and O. They seem to be low-temperature products without chondrules, planetary reheating and other alteration processes (ANDERS 1981). But, of course, all possibilities of the water exposure reactions in their parent bodies may have caused modifications of the original material. For a long time, the C1-chondrites have served as the "standard cosmic matter" from which the solar system originated.

Table 3.1. Carbonaceous chondrites: significant properties and material characterization (quantities in mass-%)

Meteorite class	significant properties and material characterization
C I	condensation temperature: ~ 300°K, no chondrules, H_2O: ~ 20%, organic matter: ~ 3.8%, layer lattice silicates, $MgSO_4 \cdot 7\,H_2O$, Fe-, Mg-, Ca-carbonates, little olivine (?)
C II	condensation temperature: ~ 400°K, chondrules: up to 50%, H_2O: ~ 13%, organic matter: ~ 2,5%, layer lattice silicates, $MgSO_4$: 7 H_2O, Fe-, Mg-, Ca- carbonates (little), olivine, pyroxene
C III	condensation temperature: ~ 430°K, metamorphic temperature: ~ 600°K, chondrules: up to 60%, H_2O: ~ 1%, organic matter: 0.5%, little layer lattice silicates, many olivine and pyroxene

The corresponding models for the origin of the solar system were based consequently on mechanisms of preplanetary condensation from a material reservoir of homogeneous and well mixed elemental and isotopic composition. But, in the last ten years and by using suitable analytical techniques it has been shown, that there was probably a nonhomogeneous material reservoir. For instance, mineral phases separated from several carbonaceous chondrites show, partially, peculiar isotopic compositions. Their origin could result from

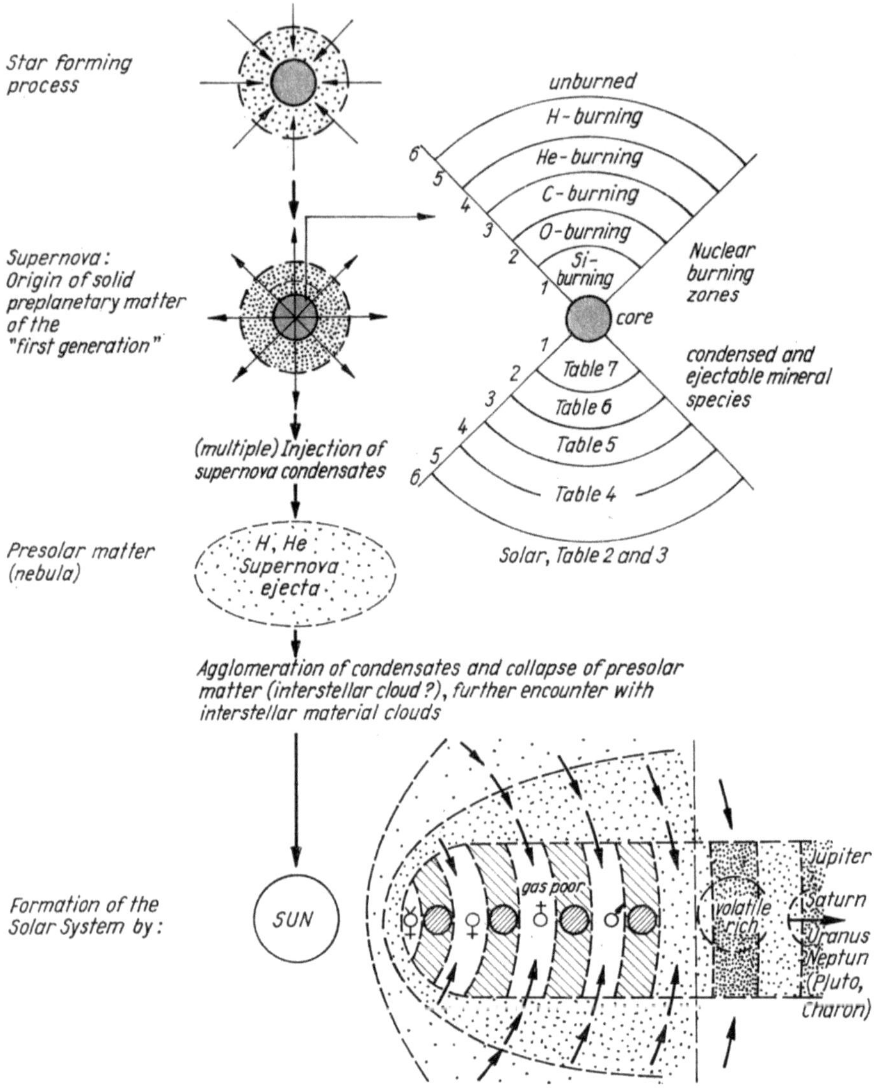

an admixture of diverse proportions of isotopes formed by a number of different presolar nucleosynthetic processes. These isotopic "anomalies" could have led to the following interpretations:

— many different stellar nucleosynthetic sites provided material for the interstellar preplanetary material reservoir and the preplanetary material cloud was not very well mixed,
— the preplanetary material in the solar system was well homogenized, but, one only finds some "survivors" in "suitable" parent bodies.

But, all the latest material results do not affect the widely accepted theoretic models of the origin of the planetary system. This process starts with condensation (homogeneous and/or heterogeneous) from a gas phase in a solar gas-dust cloud. The solid particles originated in this way grew together to planetesimals. The latter finally accreted to planetary bodies.

In Fig. 3.1a scenario is given about the origin and earliest evolution of solar system. It represents a summary of many experimental results and theoretical models. One of the most essential aspects seems to be, that the accretion of planetesimals is an intensive and short time process.

Of course, the scenario as shown in Fig. 3.1. is still a matter of speculative debate. A profound reason is, as mentioned above, that it is impossible to deduce the state of the "original" material from the actual state of planetary material. This question will remain an open problem certainly also in future. Furthermore, some exotic components of the carbonaceous chondrites (CAI's, FUN's) do not seem to be true preplanetary solids, but (later) modified planetary matter of more or less primitiveness.

In the following sections the present day situation will be presented considering aspects of preplanetary material only. Mainly, the following points of view will be in the centre of interest:

— the origin of solid preplanetary matter
(condensation and nucleation processes)

Fig. 3.1. The Formation of the Solar System: A model

a) Physical processes:

— collapse of material toward Sun's equatorial plane and formation of an imperfectly mixed material disc
— splitting of the circumsolar disc into separate rings of preplanetary matter
— origin of planets

b) Material processes (condensation, nucleation, separation, etc.):

— condensation of preplanetary Solar System gas (also from revaporized solids) to "second generation" solids
— accretion of "first and second generation" matter forming planetesimals
— modification of all formed solids by solar heating (T-Tauri-phase)
— condensation of the "rest" gaseous matter in the further cooling solar nebula.

— the nature of observable preplanetary material (circumstellar dust)
— meteorites and the nature of preplanetary matter (age, isotopic pattern, mineralogical-, petrological-, and physical properties of meteorites)
— some open problems.

3.2. The Origin of Solid Preplanetary Matter

Planetary objects will be formed by condensation, forming solid state matter (cluster), and later accretion of this matter from a presolar cloud of grains. Such a cloud consists of differently evolved matter which arises from different material forming processes and sources. Above all, that are stellar processes in connection with novae, supernovae, red giants, etc. All kinds of material formed in this way may participate in the formation of the planetary bodies also. So, the planet forming material reservoir consistend of gaseous and solid matter with different chemical and physical properties. This must be taken into account at by investigating the formation processes of planetary material.

The preplanetary material arose from events starting with condensation of a gas medium and homogeneous and/or heterogeneous nucleation of the condensed species. Further there are reactions, commonly and simultaneously, with large fractions of refractory heavy elements and compounds condensed in grains and descended from dust forming epochs in novae and supernovae.

In the following some aspects of the origin of such solids as silicates, oxides, metals, etc. will be presented.

3.2.1. Condensation Processes

Investigations and calculations regarding the condensation of solids from a solar gas phase at cosmic conditions were performed by several authors as Grossmann and Larimer (1974), Larimer and Anders (1967, 1970), Grossmann (1972), Larimer (1975), Lewis (1972), Kiesl (1979), Lattimer (1982), etc. Similar calculations were carried out for supernovae condensation processes by Lattimer, Schramm, and Grossmann (1978), Lattimer and Grossmann (1978), and Lattimer (1982).

Thermodynamic calculations allow information about equilibrium concentrations of chemical compounds or elements at a given pressure and temperature if the composition of the system is known. Today, these computations are based upon the "normal" abundance of chemical elements in the Solar System. Hence, all existing thermodynamic calculations result from relatively few data of simple compounds. Furthermore, to get more quantitative facts one needs kinetic properties additionally. But, in this connection suitable datas are not very well known. This situation is yet unsatisfactory, especially because kinetic factors (reaction velocity, etc.) have a great signifiance if the condensation took place far from the equilibrium as in the early Solar System. Uncertainties result from the fact that gaseous molecules and crystalline phases could have coexisted. The corresponding thermodynamic data are scarely known today.

Possible homogeneous and heterogeneous nucleation mechanismus are discussed by LATTIMER (1982). If condensation of a cooling gas into droplets or grains took place in the absense of preexisting condensate nuclei, homogeneous nucleation will be the dominating process. At preexisting nuclei in the cooling gas, condensation starts with heterogeneous nucleation. The heterogeneous nucleation is energetically favoured in opposite to the homogeneous ones in general. Also homogeneous nucleation is a slow process compared to the heterogeneous nucleation because of the large nucleation barrier that must be overcome for homogeneous nucleation.

It is relatively simple to describe the condensation of chemical elements from a cooling gas, consisting of hydrogen mainly (KIESL, 1979). The condensation starts if the partial pressure (p_i) equals the vapor pressure (p_0). Then the partial pressure of a chemical element in a gas of cosmic composition is equal to the product of its abundance in relation to hydrogen and is multiplied by the total pressure of the system (P).

$$P_i = \frac{A(X)}{A(H_2)} \cdot P, \tag{3.2.1}$$

$A(X)$ = normal (solar) abundance of the element X,
$A(H_2)$ = normal (solar) abundance of the hydrogen.

According to KIESL (1979) molecular hydrogen ist the most frequent hydrogen bearing compound below 2000 °K. Therefore, in the solar nebula the hydrogen pressure (P_{H_2}) is practically equal to the pressure of the system (P).

At 2000° K and if the abundance of hydrogen is assumed to be 1, the equilibrium constant (Kp) will be $10^{5.6}$ with

$$Kp = 10^{5.6} = \frac{P_{H_2}}{(P_H)^2} \tag{3.2.2}$$

and

$$P_{H_2} = 4 \cdot 10^5 \cdot (P_H)^2. \tag{3.2.3}$$

Following KIESL further, at $P_{H_2} = 1/2$ (frequency of hydrogen) becomes $P_H = 1.1 \cdot 10^{-3}$. Diminishing the temperature below 2000° K the portion of atomar hydrogen will decreases even more.

Using the Clausius-Clapeyron-function, it is possible to describe the vapor pressure of a chemical element with

$$\log p_0 = -\frac{A}{T} + B. \tag{3.2.4}$$

It is:

$$A \sim \frac{H}{2.3R}, \tag{3.2.5}$$

$$B \sim \frac{S}{2.3R}, \tag{3.2.6}$$

H = evaporation enthalpie,
S = evaporation entropie,
R = gas constant.

The partial pressure of a condensing chemical element decreases with $(1 - \alpha)$ (α = condensed proportion).

The condensation temperature is calculable then with the relation

$$\log(1 - x) = -\frac{A}{T} + B - \log P - \log \frac{A(H)}{A(H_2)}. \tag{3.2.7}$$

Assuming a total pressure of 10^{-4} atm in a gas of solar composition GROSSMANN and LARIMER (1974) calculated a condensation sequence of chemical elements which is presented in Fig. 3.2.

Here, a quickly cooling solar nebula is assumed. At such conditions, a comparatively long accretion period is supposed by KIESL (1979), which is significant at the low temperature condensation particularly. Especially if the equilibrium is reached, compounds must be considered which affected the chemical properties of the system profoundly.

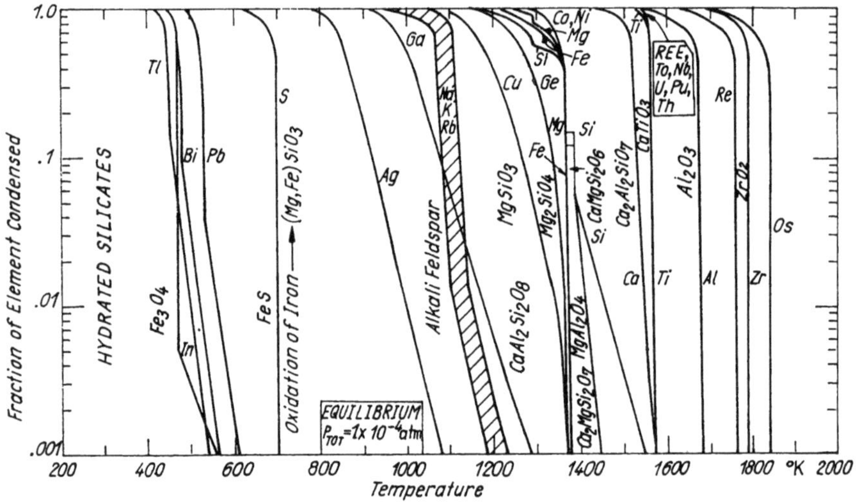

Fig. 3.2. Condensation of the elements from a gas of solar composition at 10^{-4} atm (GROSSMAN and LARIMER 1974).

As mentioned above, changes of the chemical properties affect significantly the condensation sequence. The abundances of elements and compounds will be influenced in particular. Especially the H/O- and C/O-ratios (oxidation state) control the most important abundance changes. The condensation of oxides, f.e., will be suppressed in a more reducing gas.

Using different C/O-ratios LATTIMER and GROSSMANN (1978) and LATTIMER (1982) have performed appropriate calculations. They used C/O-ratios of 0.55 and 1.2 respectively and received condensation temperatures for it. Further they calculated so called "disappearance temperatures". At these temperatures phase reactions of solids generated earlier with the gas yield a new generation of solids. Simultaneously, the "original" solids disappear (Tables 3.2. and 3.3.)

LATTIMER (1982) states, that an increase of the solar C/O value of 0.55 up to 0.9 is without succession. At greater values the condensation temperatures of refractory oxides drop. Fe condenses at C/O < 1 as an alloay, at C/O > 1 as carbide Fe_3C. Above a C/O ratio of 0.95 appear, for example, carbides,

Tables 3.2./3.3. Condensation and disappearance temperature of solids in a cooling solar nebula (LATTIMER 1982)

$P = 10^{-3}$ atm. C/O = 0.55			$P = 10^{-3}$ atm. C/O = 1.2		
mineral	condensation temperature	disappearance temperature	mineral	condensation temperature	disappearance temperature
Al_2O_3	1743		TiC	1893	1025
$CaTiO_3$	1677	1409	SiC	1742	1154
melilite	1625	1438	C	1732	871
$MgAl_2O_4$	1533	1391	Fe_3C	1463	1326
Fe–Si	1458		AlN	1390	1234
$CaMgSi_2O_6$	1438		CaS	1385	1040
Mg_2SiO_4	1433		Fe–Si	1326	
Ti_3O_5	1409	1274	Al_2O_3	1235	1229
$CaAl_2Si_2O_8$	1392	1068	$MgAl_2O_4$	1229	1077
$MgSiO_3$	1351		Mg_2SiO_4	1154	
Ti_4O_7	1274	1120	MgS	1131	1100
TiO_2	1120	774	$CaMgSi_2O_6$	1069	
Al_2SiO_5	1068		$CaAl_2Si_2O_8$	1057	1044
$NaAlSi_3O_8$	1028	780	$MgSiO_3$	1054	
$NaAlSi_2O_6$	780		Al_2SiO_5	1045	
$CaTiSiO_5$	774		TiN	1025	910
			$NaAlSi_3O_8$	961	780
			Ti_4O_7	910	862
			TiO_2	862	770
			$NaAlSi_2O_6$	780	
			$CaTiSiO_5$	770	

sulfides, and nitrides and Ti, C, and Si are the most refractory phases. According to LATTIMER (1982) only half of the Si and less than half of the C condense at high temperatures in the case of $C/O \lesssim 2$. The temperatures of complete condensation of the elements Al, Si, Ca, and Mg decrease by 350° K, 250° K, 200° K, and 200° K, respectively, in comparison with the case of $C/O < 1$. It is also pointed out, that the Fe condensation temperature is scarcely affected by changes of the C/O ratio.

As mentioned above, also the H/O ratio is able to alter the oxidation state of a solar gas. LATTIMER and GROSSMANN (1981) have carried out such calculations. It has been indicated, that oxides condense at higher temperatures, if the H/O ratio is low. Further, metallic Fe appears mostly in the form of Fe-oxide. The temperature of complete condensation of Al, Ca, Mg, Na, and Si is not affected. For the case of condensation of a solar type gas. LATTIMER

Tables 3.4/3.7. Condensation products and condensation and disappearance temperatures of the most abundant elements (LATTIMER 1982)

Table 3.4. (H), (He) Zones
($P_I = 10^{-8}$ atm.)

a) C/O = 0.55			b) C/O + 1.5		
mineral	condensation temperature	disappearance temperature	mineral	condensation temperature	disappearance temperature
Al_2O_3	1597	1390	C	1950	
melilite	1467	1332	TiC	1550	747
$MgAl_2O_4$	1390	1340	SiC	1308	783
$CaTiO_3$	1370	1306	Fe	1053	
$CaAl_2Si_2O_8$	1340	1328	AlN	1035	890
$CaMgSi_2O_6$	1332		CaS	1020	610
$MgAl_2O_4$	1331	1281	Al_2O_3	990	878
Mg_2SiO_4	1320		$MgAl_2O_4$	878	766
$MgTi_2O_5$	1306	1255	Mg_2SiO_4	829	746
$CaAl_2Si_2O_8$	1281	1071	Al_2SiO_5	783	555
TiO_2	1255	1120	$MgSiO_3$	753	
$MgSiO_3$	1251		MgS	752	695
$FeTiO_3$	1120		TiN	747	650
FeO	1087		SiO_2	729	710
$Al_6Si_2O_{13}$	1071	1068	$NaAlSi_2O_5$	715	
Al_2SiO_5	1068		FeS	710	
$NaAlSi_3O_8$	807	775	Mg_2SiO_4	695	
FeS	786		Ti_4O_7	650	635
$NaAlSi_2O_6$	775		$CaTiSiO_5$	635	
			$CaMgSi_2O_6$	610	515
			$Ca_3Al_2Si_3O_{12}$	555	
			$CaSiO_3$	515	

(1982) concluded altogether, that changing factors of 2—3 of the major element abundances (Al, Ca, Fe, Mg, Na, S, Si, Ti) does not apprecially affect the oxidation state. It will be expected furthermore, that under the described conditions the chemical composition is very similar to the solar one in various stellar environments (novae, red giants, planetary nebulae, protosolar nebulae). The essential factor is the C/O ratio. Unlike this, it is possible, that there is an other chemical situation in supernova ejecta. Following LATTIMER, SCHRAMM and GROSSMANN (1978) and LATTIMER and GROSSMANN (1978) there will be specific chemical compounds produced in the individual layers of the supernovae (Fig. 3.1.). The matter in these layers will be burned at different energy levels during the explosion of the supernova. So, a layered construction is assumed with zones characterized by Si, O, C, He, H, and "unburned" from the center outward. The condensation and disappearance temperatures of the most abundant elements (or compounds) formed by specific processes in each layer are listed in Tables 3.4.—3.7.

LATTIMER (1982) supposes that in the outermost supernova layers there are nearly solar abundances. In the next inner layer (H, He; Table 3.4.) Fe condenses as FeO and Fe_3O_4. FeO probably is very stable, therefore solid solutions with Mg-silicates should appear.

Going to the next inner layer (C; Table 3.5.) the abundances of C, O, Na, Mg, and Al are increased at the expense of He and N relative to solar abundances. Ca-phases like melilite are clearly less abundant compared with Al-

Table 3.5. (C) Zone ($P_I = 10^{-8}$ atm.)

a) C/O = 0.5			b) C/O = 2.0		
mineral	condensation temperature	disappearance temperature	mineral	condensation temperature	disappearance temperature
Al_2O_3	1830	1617	C	2150	
$MgAl_2O_4$	1617		TiC	1530	792
melilite	1595	1495	SiC	1470	920
Mg_2SiO_8	1522		Al_2O_3	1120	990
$Ca_3MgSi_2O_8$	1495		Fe_3C	1095	1055
MgO	1460		CaS	1065	
$CaTiO_3$	1451		Fe	1055	
FeO	1208		$MgAl_2O_4$	990	
Na_2SiO_3	760		Mg_2SiO_4	935	
FeS	740		MgS	918	703
Ca_2SiO_4	690		MgO	905	
			Mg_2TiO_4	793	670
			Na_2S	703	
			$CaTiO_3$	670	
			Na_2SiO_3	652	
			FeS	550	

Table 3.6. (O) Zone ($P_I = 10^{-8}$ atm.)

a) O-rich

mineral	condensation temperature	disappearance temperature
melilite	1655	1340
Ca_2SiO_4	1610	1370
$CaTiO_3$	1480	1320
$CaSiO_3$	1394	865
$CaAl_2Si_2O_8$	1340	820
$CaTiSiO_5$	1320	1040
SiO_2	1287	
$CaMgSi_2O_6$	1140	807
Fe	1110	
Ti_2O_3	1040	965
TiO	965	
Si	915	
CaS	880	
Al_2SiO_5	820	
$MgSiO_3$	807	750
MgS	750	
$NaAlSi_2O_6$	690	
SiS_2	600	

b) S, Si-rich

mineral	condensation temperature	desappearance temperature
Si	1390	
TiO	1169	
CaS	1130	
Fe	1110	
Al_2O_3	1070	1060
melilite	1060	1020
$Ca_3Al_2Si_3O_{12}$	1020	940
$CaSiO_3$	1015	880
$CaAl_2Si_2O_8$	940	830
$CaMgSi_2O_6$	913	810
SiO_2	880	
Al_2SiO_5	830	
$MgSiO_3$	810	760
MgS	760	
$NaAlSi_2O_6$	690	
SiS_2	610	

Table 3.7. (Si) Zone

Fractional number abundances

Ca	0.03
Fe	0.75
S	0.10
Si	0.11
Ti	5×10^{-4}

a) $P = 10^{-4}$ atm.

mineral	condensation temperature
Fe	1820
Si	1670
Ti	1620
CaS	1490

b) $P = 10^{-12}$ atm.

mineral	condensation temperature
Fe	1089
Si	1083
Ti	1075
CaS	950

phases like corundum or spinel. That is in contrast also to solar relations. Because the Mg/Si-ratio is relatively high, Mg_2SiO_4 (forsterite) is the main Si-bearing phase. Si condenses at higher temperatures and also earlier than Mg. Ti appears in $CaTiO_3$ at 1400° K. At higher temperatures Al, Ca, and Si condense what is also in contrast to solar relations. If more C plays a role as it does in the solar case, carbides and sulfides of the elements Si, Ti, and Fe condense at first. But, the complete condensation of Al, Ca, Si, and Mg remains in te same order as in the C-poorer model.

The next inner layer is shown in Table 3.6. (O). It contains large abundances of Si, S, O, and Ca. In the O-rich case, a large Ca/Al ratio favours Ca-rich compounds at the beginning of the condensation. This causes simultaneously the absence of oxides and related compounds. The Si/Mg ratio is very high, therefore Mg-silicates are suppressed. Because the Si/O ratio is high in comparison to solar relations, it is also possible that Si condenses elementally if all O is bound in SiO_2. Together with other effects (Si, Fe-alloy) Si might appear at 200° K higher temperatures. The Si-rich model is "reducing". The elements C and N are not abundant, therefore nitrides and carbides do not belong to the first condensates. There rather appear metallic Si and Fe, CaS and TiO. In such environment oxides and silicates are late condensates. The total solidification in the O-layer is reached for Ca and Ti at 1100° K, for Al at 1050° K, for O at 900° K, for Mg at 800° K, for Na at 675° K, and for S and Si at 575° K. The supernova "core", Lattimer's Si-zone, is characterized by a condensation sequence shown in Table 3.7. It is likely that, metals condense earlier than Ca, Fe and Si-sulfides. Eventually Ti-sulfides also condense in this area.

3.2.2. Nucleation Processes

The origin of solid bodies begins with the nucleation of the corresponding material clusters. Unfortunately, there is neither an exact theoretical modelling nor are there any suitable experimental data available. Therefore, today it is not possible to simulate truely astrophysical conditions by laboratory experiments. Nevertheless, Donn (1984) tried an experimental approach, investigating refractory materials. He started from the following conditions, supported to be natural:

There is approximately one percent (of mass) of interstellar grains in the Milky Way. These grains, including ice-coated ones (see also Greenberg-particles), have diameters between several tens and one hundred nm (Spitzer, 1978), lifetimes in the order of 10^7 years (Salpeter, 1979), and a production rate of $10^{28}\ gs^{-1}$. Furthermore, there is a low total density and also a low density of condensible species. Although there is no uniform chemical composition anywhere, Donn (1984) also took into account an "averaged" elemental distribution which is representative in a large quantity of circumstellar shells. Suitable objects are O-rich stars with element abundances as shown in Table 3.8.

Using the collision kinetics for such a system Donn (1984) has shown a comparison between laboratory parameters and the "real" conditions in circumstellar shells (Table 3.9.).

Table 3.8. Abundance of 11 major elements around O-rich stars (TRIMBLE 1975)

Element	Abundance	Element	Abundance
H	$3 \cdot 10^{10}$	Fe	$8 \cdot 10^5$
O	$2 \cdot 10^7$	S	$5 \cdot 10^5$
C	$1 \cdot 10^7$	Ca	$7 \cdot 10^4$
N	$4 \cdot 10^6$	Na	$6 \cdot 10^4$
Si	$1 \cdot 10^6$	Ni	$5 \cdot 10^4$
Mg	$1 \cdot 10^6$		

Table 3.9. Kinetics of collisions (from DONN 1984)

	1	2	3
a) Temperature (°K)	300	750—1000	1000
b) Total density (cm^{-3})	10^{19}	10^{17}—10^{18}	10^8—10^{11}
c) Time between collisions (s)	10^{-8}	10^{-7}—10^{-6}	10^3—10^0
d) Condensible density (cm^{-3})	10^{17}	10^{13}—10^{16}	10^4—10^7
e) Collision time condensibles (s)	10^{-6}	10^{-6}—10^{-3}	10^3—10^6

a) temperature of the system
b) density of the total gas (mainly hydrogen)
c) time between collisions in the total gas
d, e) like b, c for condensible species

1 — representative values for nucleation experiments
2 — characteristic values of Donn's experiments on refractory condensates
3 — values of circumstellar shells

The condensible species are mainly SiO, Fe, Mg, S, Al, Ca, and Na. Suitable collisions forming grains occur in relatively long time intervals, after DONN (1984) in a range of hours up to days. Consequently, the originating clusters have a considerable life time.

Experiments on the formation of refractory grains using vapour phase condensation were carried out f. e. by DAY and DONN (1978), STEPHENS and RUSSEL (1979), STEPHENS and BAUER (1981), NUTH and DONN (1983, 1983b, 1984). It has been demonstrated, that the investigated gas phases resemble the observed cosmic condensates.

Condensation experiments with SiO-vapour yielded amorphous Si_2O_3, if the SiO vapour condensates in an argon or hydrogen environment. This Si_2O_3 disproportionates relatively easily to amorphous quartz plus elemental silicon. Further, Si_2O_3 has a clear infrared feature at 11.5 μ. The same signature is observable in shells of relatively cool oxygen-rich stars. NUTH and DONN

(1984) showed also, that a transition of Si_2O_3 in amorphous SiO_2 (by annealing) is more probable than a transition to an amorphous Mg—SiO condensate.

Besides the experiments, theoretical evaluations have shown that the transformation reaction from Si_2O_3 into amorphous SiO_2 appears to be a unimolecular disproportionation with the possibility that such clusters, condensed in cool circumstellar shells, could be transported unaltered into the interstellar matter. Of course, other silicate clusters also can be formed.

They all were incorporated in the presolar material cloud. There, these clusters, grains, etc. served as condensation and also nucleation centers (heterogeneous nucleation). The formation and the growth of solid particles by means of such centers is favoured energetically in contrast to f. e. homogeneous nucleation. Consequently, the generation of the "first generation" of planetary solids might be a quick process. But, the efficiency of the condensation and nucleation centers depends on the degree of supersaturation and the position in the Solar System. In the outer part of the nebula there is always a large number of grains which can act as such centers. In the inner solar system, at higher temperatures, preexisting grains will become evaporated, eventually completely. Under these conditions, the condensation and nucleation processes could be dominated by the thermodynamic properties of a very hot solar nebula. In this region the density and supersaturation was very high. Therefore, as the temperature sank deep enough condensation may have occured again in a very short time interval. But, this is possible only by a rapid transport of the energy out of this region of the Solar System.

The properties of the above mentioned silicatic clusters or grains are not strictly classificable. Spectroscopic evidences from interstellar dust suggest that these particles are, at least in parts, amorpheous. In contrast, silicates which are observable directly (f.e. interplanetary dust particles) are generally crystalline rather than amorphous. All together, it seems that generation, evolution, and destruction (transformation, etc.) of interstellar (presolar) grains is a continuous process. The grains are always undergoing any change. Therefore, a great ensemble of silicate particles with very different properties apparently participated in the formation of planetary parent bodies.

A possible intermediate stage in the growth to greater planetary bodies might have been the so-called "Greenberg"-particles (Greenberg, 1984). They consist of silicate core enveloped in different organic compounds, molecules, radicals, etc. Their existence results from observations and laboratory and theoretical investigations. Greenberg's particles should be formed in different "evolved" interstellar clouds. In low density diffuse clouds the grains have a silicate core and an envelope consisting of refractory organic matter. Grains in more dense molecular clouds have an additional envelope of lighter molecules and radicals. Ultimately around protostellar objects at the final stage of collapse (leading to a star) grains will be formed which have accreted all remaining organic species. Further it is shown that H_2O is more abundant in solid grains than in the gas phase of the above described molecular clouds or of clouds around protostellar objects. Thus, the Greenberg-particles could represent a "second" generation of material which accretes later to planetesimals.

3.3. The Nature of Observable Preplanetary Material: Circumstellar Dust

DORSCHNER (1971) suggested that a classification of three dust types is conceivable:

a) Dust grains formed in normal interstellar gas clouds. The chemical composition of these grains should approximately reflect the mean cosmic abundance of the chemical elements.
b) Dust grains formed originally in stellar atmospheres (but also in supernovae, novae, ...). These solids consist of refractory material and must show a related chemical composition
c) Dust grains formed in circumstellar space during and after the formation of a star.

Today it is an obvious observational fact that condensation processes take place in star atmospheres, circumstellar nebulas, etc. Infrared investigations confirm the production of dusty matter (elements, oxides, silicates) by the above mentioned process. Other compounds as radicals and molecules like OH, H_2O, NH_3, H_2CO etc. were detected in interstellar star forming regions. Between these species interactions occur. These chemical reactions are not only effective during most hot stages of the star formation and evolution, but also a long time after the origin of the new star.

After DORSCHNER (1971), material producing processes in circumstellar space can contribute significantly to the dust in interstellar space. The resulting material is not homogeneously distributed throughout the Milky Way. This can be deduced from obscuration and reddening of star light (SAVAGE and MATHIS 1979). Observations of expanding star atmospheres, ejected material from cool giants, or violently ejected matter from novae show an infrared continuum similar to that produced by particles with temperatures of about 1000° K.

Very interesting related results were provided especially by the observation of the luminosity of Nova Vulpeculia 19 at visible and infrared wavelenths. There was a sudden drop in visual luminosity and, nearly simultaneously, a sudden rise in infrared emission. This is a strong evidence for the formation of solid particles in the expanding ejecta. Simultaneously with the obscuration of visible light by the dust particles, the thermal infrared radiation of the solids increased. This effect was observed fifty days after the outburst, indicating the time interval up to the formation of solid particles.

Having in mind this complicated scenario it will be very difficult to find survivors from this period. It is more probable that all components are altered. But this is a yet open question.

3.4. Meteorites and the Nature of Preplanetary Matter

The ordinary chondrites, particularly the carbonaceous ones, are the most primitive objects (eventually containing remnants of preplanetary matter) which are up to day recognized as the best, most complete, and most coherent source of data when investigating the composition of the material out of which the Solar System is formed.

As mentioned above, it was assumed originally that the preplanetary material of the Solar System was a well-homogenized mixture of matter. But, since the discovery and investigation of both the high-temperature inclusions of carbonaceous chondrites (f. e. in the Allende C3-type chondrite) and the isotopic anomalies of chemical elements like O, Mg, Ca, Ti, Sr, Ag, Ba, Nd, Sm, etc. a homogenized material source can not longer be assumed. Further, it is probable that the constituents of the Solar nebula were formed by different stellar processes and not from one source alone. Consequently, there is no unique way to determine the history of the Solar System from investigations of only a few meteorites.

Nevertheless, in the following sections some properties of several preplanetary material forming events will be discussed. This is based mainly on investigation of chondrites.

3.4.1. The Age of Meteorites

First of all, a suitable approach to identify the oldest solid bodies and/or material of the Solar System is the determination of their age. Therefore, many experimental investigations were carried out in the past in order to determine the radiometric age of planetary matter. Samples available for establishing a time frame for planetary origin and evolution comes from the Earth, the Moon and from meteorites. For answering the formulated questions, rocky materials of planets are unsuitable, because subsequent processes of the planetary evolution can reset the radiometric clocks. Therefore it is more helpful to analyse the most primitive solids of the Solar System which are available, namely the chondritic meteorites.

Today it is generally accepted, that the Solar System originated about $4.6 \cdot 10^9$ years ago. If the chondritic meteorites represent, at least in parts, remnants of this epoch they must show a comparable age. Furthermore, if there is surviving presolar dust relatively unaltered in primitive meteoritic material, this dust component might still be older. Suitable candidates could be the Ca—Al-rich inclusions (CAI's) from the Allende carbonaceous chondrite.

Results of the determination of radiometric ages of the individual meteorites or meteorite constituents were compiled by Kirsten (1978) and are shown in Tables 3.10 and 3.11.

In comparison to Kirsten (1978) also some results for whole meteorites (classes) given by Allegre (1982) are shown in Table 3.12.

Table 3.10. Radiometric ages of meteorites (from KIRSTEN 1978)

Type	Meteorite	Method	Age ($\times 10^9$ years)	Remarks
C 5	Allende	Pb—Pb internal isochron	4.565 ± 0.004	inclusions
		Pb—Pb internal isochron	4.553 ± 0.004	
		Pb—Pb ion probe	4.53—4.64	no primordial Pb
		Pb—Pb upper intercept	4.57 ± 0.02	
		Th—^{208}Pb internal isochron	4.565	
C 4	Karoonda	Ar—Ar plateau	4.60 ± 0.10	
L 4	Bjurböle	Ar—Ar plateau	4.60 ± 0.03	
L 4	Saratov	Ar—Ar plateau	~ 4.60	
L 6	Peace River	Rb—Sr internal isochron	4.56 ± 0.03	one subsample deviates
LL 4	Soko Banja	Rb—Sr internal isochron	~ 4.50	
LL 5	Krähenberg	Rb—Sr internal isochron	4.70 ± 0.02 (?)	
LL 5	Olivenza	Rb—Sr internal isochron	4.63 ± 0.16	
		Ar—Ar plateau	4.50 ± 0.03	
LL 6	St. Mesmin	Ar—Ar plateau	4.57 ± 0.05	Xenolite
LL 6	St. Severin	Rb—Sr internal isochron	4.55 ± 0.09	
		Pb—Pb internal isochron	4.53 ± 0.02	
		Ar—Ar plateau	4.50 ± 0.03	
L	Barwell	Ar—Ar plateau	~ 4.53	
H 4	Forest Vale	Total K—Ar	4.61 ± 0.1	
		Total U/Th—^{4}He	4.7 ± 0.2	
H 4	Ochansk	Ar—Ar plateau	~ 4.57	
H 5	Allegan	Ar—Ar plateau	4.60 ± 0.1	
H 5	Guarena	Ar—Ar plateau	4.50 ± 0.03	
		Rb—Sr internal isochron	4.56 ± 0.08	
H 6	Mount Browne	Ar—Ar plateau	4.58 ± 0.05	
H 6	Queens Mercy	Ar—Ar plateau	4.58 ± 0.05	
H	Nadiabondi	Total K—Ar	4.56 ± 0.1	
		Total U/Th—^{4}He	4.55 ± 0.2	
H	Utzendorf	Total K—Ar	4.5 ± 0.1	
		Total U/Th—^{4}He	4.4 ± 0.2	
E 4	Adhi Kot	Total K—Ar	4.7 ± 0.1	
		Total U/Th—^{4}He	4.7 ± 0.2	

Table 3.10. (continued-1)

Type	Meteorite	Method	Age ($\times 10^9$ years)	Remarks
E 4	Indarch	Rb—Sr internal isochron	4.56 ± 0.15	
Au	Norton County	Rb—Sr internal isochron	4.70 ± 0.10	possibly not in equilibrium
Di	Tatahouine	Rb—Sr internal isochron	~ 4.55	
An	Angra dos Reis	Pb—Pb internal isochron	4.548 ± 0.002	
		U—^{206}Pb internal isochron	4.61 ± 0.07	
		Th—^{208}Pb internal isochron	4.57 ± 0.10	
		Sm—Nd internal isochron	~ 4.42	
Eu	Ibitira	Rb—Sr internal isochron	4.53 ± 0.10	
	Juvinas	Rb—Sr internal isochron	4.60 ± 0.07	
		Sm—Nd internal isochron	4.56 ± 0.08	
Ho	Kapoeta	Rb—Sr internal isochron	4.54 ± 0.12	Clast
Iron	Mundrabilla	Ar—Ar plateau	4.62 ± 0.02	Olivine inclusions
		Ar—Ar plateau	4.58 ± 0.02	Plagioclase inclusions
Iron IA	Toluca	Total K—Ar	4.63 ± 0.1	Silicate inclusions
Iron IB	Four Corners	Total K—Ar	4.58 ± 0.1	Silicate inclusions
Iron IIE	Colomera	Rb—Sr internal isochron	4.61 ± 0.04	Silicate inclusions

C: Carbonaceous chondrite, L: low iron chondrites (hyperstene chondrite), LL: low iron, low metal chondrites (bronzite chondrite), E: enstatite chondrite, Au: Aubrite (Ca-poor achondrite), Di: Diogenite (Ca-poor achondrite), An: Angrite (anamalous Ca-rich achondrite), Eu: Eucrite (Ca-rich achondrite), Ho: Howardite (Ca-rich achondrite), Type 3—6: degree of metamorphism, IA, IB, IIE: iron meteorite subgroups

Minster and Allegre (1979a, 1979b) established a relative error of $\pm 20 \cdot 10^6$ years at the ^{87}Rb—^{87}Sr dating method for H, E, and LL-type condrites. In general, an average "age line" of about $4.5 \cdot 10^9$ years was found. All of the investigated ordinary chondrites fit to this value. That will also be confirmed by the results using other dating methods. Allegre (1982) gives a best age of $4.503 \pm 0.015 \cdot 10^9$ years, taking into consideration all available datas of chondrite whole-rock ages.

Table 3.11. Whole rock isochron ages of meteorite groups (from KIRSTEN 1978)

Meteorite group	number	Method	Age ($\times 10^9$ years)	Remarks
L-Chondrites	6	Rb–Sr	4.54 ± 0.12	Unshocked falls
LL-Chondrites	10	Rb–Sr	4.56 ± 0.15	Falls only
H-Chondrites	12	Rb–Sr	4.69 ± 0.14	Falls only
H-Chondrites	3	$^{207}Pb-^{206}Pb$	4.55 ± 0.03	
C-Chondrites	5	Rb–Sr	4.46 ± 0.35	
C-Chondrites	5	Rb–Sr	~ 4.69	Other C-Chondrite deviate
E-Chondrites	8	Rb–Sr	4.54 ± 0.13	Falls only
C + L + H-Chondrites	16	$^{207}Pb-^{206}Pb$	4.505 ± 0.008	
C + L + H-Chondrites		$^{207}Pb-^{206}Pb$	~ 4.57	
Eucrites	7	Rb–Sr	4.39 ± 0.26	
Eucrites	4	$^{207}Pb-^{206}Pb$	4.59 ± 0.02	
Irons (Silicate inclusions)	5	Rb–Sr	4.6 ± 0.2	
Irons (metal)	14	Re–Os	4.0 ± 0.8	
Ch + Ach + Ions	4	$^{207}Pb-^{206}Pb$	~ 4.5	
Ch + Ach + Irons	33	$^{207}Pb-^{206}Pb$	4.50 ± 0.03	
Ch + Ach + Irons	15	$^{207}Pb-^{206}Pb$	4.50 ± 0.05	

(Ages are adapted for U – decay constants)

Table 3.12. ^{87}Rb-^{87}Sr ages of meteorite groups (ALLEGRE 1982)

Meteorite types	Age (years $\times 10^9$)
Whole chondrites	4.498 ± 0.015
H-chondrites	4.518 ± 0.026
LL-chondrites	4.486 ± 0.020
E-chondrites	4.508 ± 0.037
H3-chondrites (Tieschitz)	4.53 ± 0.06
LL4-chondrites (Soko Banja)	4.452 ± 0.020
LL6-chondrites (Saint Severin)	4.51 ± 0.15
E4-chondrites (Indarch)	4.393 ± 0.045
Whole achondrites	4.47 ± 0.10
Basaltic achondrites (Juvinas)	4.50 ± 0.1
Enstatite achondrites (Norton County)	4.38 ± 0.04

Because the age datas of all chondrites are not very different from each other it is possible to assume that the isotope pattern at the formation of these meteorites were relatively homogeneous. The ages of the Allende inclusions are also close to that of the chondrite ones. This indicates a quite limited formation time interval for the origin of different meteorite classes and constituents. The same results were reached by EVENSEN et al. (1978) who investigated chondrules of chondrites with Rb—Sr dating methods. They also found age values, which are comparable with $4.503 \cdot 10^9$ years.

Of course, the presented exactness of the dating methods is not sufficient to establish a very close age scale. Different methods very often yield age differences up to $100-200 \cdot 10^6$ years.

3.4.2. The Age of Carbonaceous Chondrites

Compared with other chondrites it is relatively difficult to get age data of carbonaceous chondrites (MITTELFEHLDT and WETHERHILL, 1979). Nature and particularly mechanical properties of carbonaceous chondrites are in that manner that terrestrial contamination is almost inevitable. But, as shown by ALLEGRE (1982), several carbonaceous chondrites fit the $^{87}Rb-^{87}SR$ isochron for chondrites. That suggests the same or at least a very similar age of carbonaceous and ordinary chondrites.

Using the U—Pb dating method. TATSUMOTO et. al. (1976) received an age below $4.4 \cdot 10^9$ years for the Orgueil matrix. Further they obtained for the white inclusions of the Allende meteorite ages up to $4.55 \cdot 10^9$ years. Basing upon the U—Pb fractionation it was pointed out that the matrix appears to be much younger than the white inclusions. Thus, all results seem to indicate, that the volatile-rich dark matrix is the most young material of the carbonaceous chondrites. Chondrules show the same age as is known from ordinary chondrites. The white inclusions, as contained in the Allende meteorite, are the oldest ones. Whether the Allende inclusions predate the solar system is still not quite clear. After GRAY et. al. (1973), the refractory inclusions are only a few million years older than the solar system. Opposite to this, JESSBERGER (1984) obtained other results. By analysing a coarse grained refractory inclusion of the Allende meteorite by investigating the Ar isotopic composition he found an apparent $^{40}Ar-^{39}Ar$ isochron age of $4.85 \cdot 10^9$ years. A re-analysis of another sample of this inclusion yielded the same result. This is a substantially greater age than estimated for the Solar System. The knowledge of the age of such inclusions is also essential, because this gives the possibility to decide where this material originated. If this matter is only a few million years older than the Solar System, then it should belong to the Solar System very probably. The process of birth could be a supernova which predates the Solar System only few million years. But, if a significant older material is found, its origin might have taken place anywhere in the interstellar space also.

3.4.3. The Age of Differentiated Meteorites

It is accepted in general that the differentiated meteorites derive from parent bodies with an own thermal history, which produced all of the chemical, mineralogical, and petrological properties of these meteorites. MANHES and

ALLEGRE (1980), using the U—Pb method, got an age of the Juvinas meteorite (basaltic chondrite) of $4.54 \cdot 10^9$ years. That means that magmatic events took place already in a very early period. Results of other age determinations confirm the crystallization at a comparable age (Nuevo Laredo at $4.55 \cdot 10^9$ years by TATSUMOTO et al. 1973, Angara dos Reis at $4.555 \cdot 10^9$ years by WASSERBURG et al. 1977). These age values were corroborated by using other methods $^{147}Sm-^{143}Nd$, $^{39}Ar-^{40}Ar$, etc.).

According to LUCK et al. (1980), iron meteorites and chondrites show a common isochron at $4.52 \cdot 10^9$ years ($^{187}Re-^{187}Os$ method), confirming very early magmatic and differentiation events. Taking into account all present experimental uncertainties and also the unknown primary age for the differentiated parent bodies of the iron meteorites, achondrites, etc. it seems that the origin as well of primitive planetesimals (eventually like carbonaceous chondrites) as well as of larger parent bodies producing suitable magmatic and differentiation events happened in an extraordinary short time interval. That means that accretion to at least planetesimal like bodies and the following heating of meteorite parent bodies to suitable temperatures were very fast processes.

3.4.4. The Isotopic Pattern of Meteorites

Up to the work of CLAYTON et al. (1973) models prevailed which based on the assumption that the Solar System was formed from a homogeneous material reservoir. CLAYTON et al. (1973) discovered a significant oxygen isotope anomaly in some constituents of the Allende meteorite. Subsequently many other isotopic anomalies have been found also in other meteorites. Today it is an accepted fact that primitive meteorites, above all carbonaceous and chondritic meteorites, contain isotopically anomalous matter. Such anomalies represent deviations from the bulk Solar System isotopic pattern of some elements. On this occasion it is significant that these anomalies are not explainable by processes which affect the isotopic abundances within meteorites. Normally the isotope content changes by mass depending on fractionation processes like diffusion, chemical reactions, by decay of radionuclides and spallation reactions. Now the discovered isotopic anomalies seem to show that the exotic compositions originate from the admixture of isotopes formed in specific presolar nucleosynthetic sites. These admixtures would then give a reference to the stellar and interstellar environment of the protosolar nebula. Their presence in only some (not all) meteorites or in some constituents demonstrates the inhomogeneous nature of the protosolar nebula as already mentioned.

LAVRUKHINA (1980) subdivides all isotope anomalies found in meteorites into two groups:

1. D, ^{10}Be, ^{20}Ne, ^{26}Mg, ^{107}Ag, and ^{202}Hg are believed to originate by nuclear reactions of high energy protons. The preplanetary solids were irradiated by protons which came from the Sun before it reached the main sequence. But, the real irradiation mechanism is not quite clear (see sect. 3.4.5.).

2. ^{13}C, ^{14}N, ^{15}N, ^{16}O, ^{120}Te, ^{122}Te, ^{123}Te, ^{130}Te, ^{86}Kr, ^{124}Xe, ^{126}Xe, ^{134}Xe, ^{136}Xe, ^{130}Xe, ^{29}Si, ^{35}S, and U. Further isotope anomalies were found at the elements Sr, Ba, Nd, and Sm in the refractory inclusions of carbonaceous chondrites (Allende, etc.). After Lavrukhina (1980) these anomalies can not be caused by high energy protons. They are rather explainable by element and isotope fractionation processes. In the case of the heavy Xenon isotopes the fission of extinct superheavy elements is proposed. Finally, there is the conclusion that the refractory inclusions derive from expanding ejecta of supernova explosion(s) and therefore the observed anomalies were established during a supernova. There are three mechanisms conceivable altogether.

The isotopic anomalies might have been caused by:

- processes far from the solar system inside supernovae or red giants,
- nearby Solar System supernovae which caused the collapse of the presolar nebula,
- supernova processes from which the solid Solar System material might have been derived directly.

The oxygen anomaly:

As has been shown by Clayton et al. (1973), the oxygen isotope ratio correspond to that, as given in Figure 3.3.

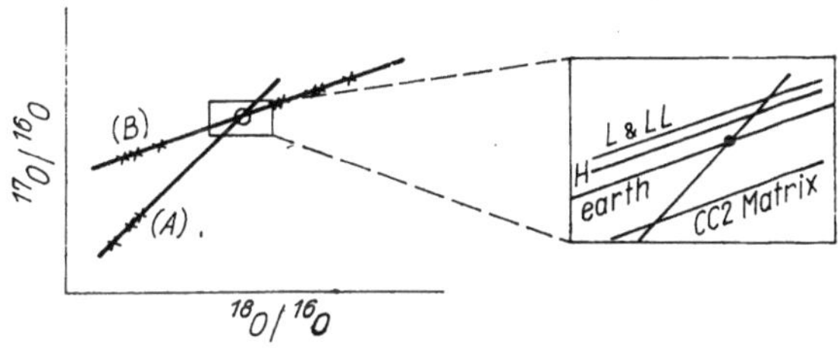

Fig. 3.3. Ratio of $^{17}O/^{16}O$ to $^{18}O/^{16}O$ in terrestrial and lunar rocks, ordinary chondrites and C2- and C3-chondrites (line *B*) and in refractory inclusions of the Allende chondrite (line *A*).
(Bibring and Langevin 1982)

Line *B* has a slope of 1/2, line *A* a slope of ~ 1. Clayton et al. (1976) assume that the slope of 1/2 represents a reservoir of single isotope pattern submitting from a mass dependent fractionation process. In contrast, the oxygen isotope ratio in refractory inclusions is interpreted as a mixing line between a "normal" oxygen component and a "pure" original ^{16}O component. The oxygen isotope variations in both terrestrial and extraterrestrial material are shown in Figure 3.4.

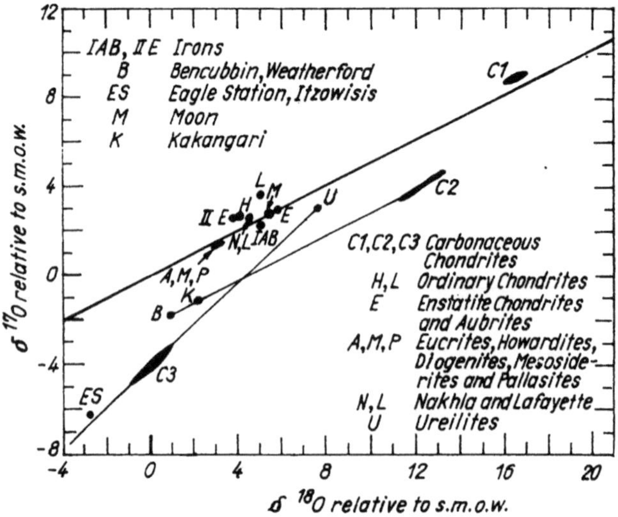

Fig. 3.4. Three-isotop plot for oxygen in $^{17}O/^{16}O$ versus $^{18}O/^{16}O$, both relative to s.m.o.w.(+) after CLAYTON (1981). All terrestrial materials lie on the heavy line with with slope 1/2. The ^{16}O-mixing line (dashed line) represents results from anhydrous minerals of C2- and C3-chondrites. The matrix material of C2-chondrites lies in the range marked with C2.

As has been shown by CLAYTON (1981), the oxygen isotope values of meteorites do not lie on the terrestrial fractionation line. They are on other lines, but also with slope ~ 1/2. But it is impossible to jump from one line to the next only by mass fractionation processes. This effect is explainable possibly only in terms of different nuclear origin. That means that the diverse meteorites contain presolar grains condensed in different material reservoirs. The largest anomalies appear in the refractory (white) inclusions and other anhydrous phases of the C2- and C3-carbonaceous chondrites. These oxygen isotope ratios define a line with slope ~ 1. The $^{17}O/^{18}O$ ratio of these samples is

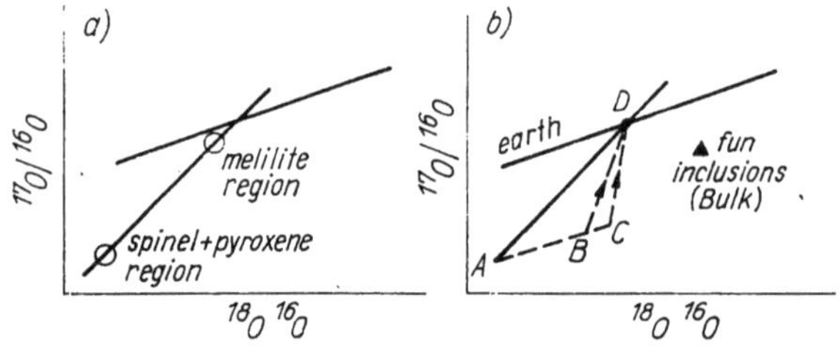

Fig. 3.5a/b. Ratio of $^{17}O/^{16}O$ to $^{18}O/^{16}O$ in terrestrial rocks and refractory inclusions and the so called FUN inclusions of meteorites after BIBRING and LANGEVIN/1982.

nearly constant and the variations correspond to ^{16}O abundance differences up to 5%. This fact has been presented by BIBRING and LANGEVIN (1982) in a more clear sketch (Figures 3.5.a and 3.5.b).

As mentioned above the greatest ^{16}O anomaly reaches up to 5%. This result has been obtained by oxygen isotope investigations of separated minerals of the refractory inclusions. Especially spinel ($MgAl_2O_4$) and fassaite (Ca(Mg, Al, Ti) $(Si, Al)_2O_6$) show an extreme ^{16}O anomaly. On the contrary, the lowest ^{16}O anomaly shows melilite (Ca, $AlSiO_2$).

CLAYTON (1977) explains the difference of the oxygen isotope abundances as follows. Originally, all condensed refractory phases exhibit the same anomaly of 5%. The condensates were heated at a later thermal "reworking". Only melilite reached a partially molten state during this process whereby an oxygen exchange took place with surrounding gas. On this way, melilite lost a little ^{16}O. But, according to BIBRING and LANGEVIN (1982), a simple mixing process is not able to explain such anomalies. They pointed out, that the presolar grains were influenced and altered rather by different processes. This is supported by the discovery of peculiar ^{16}O anomalies in the so called FUN inclusions (FUN = Fractionation and Unknown Nuclear effects). These values don't lie on the mixing line $A - D$ in Figure 3.5b. The peculiar ^{16}O anomalies might be established by a process as has been shown by CLAYTON and MAYEDA (1977). They assume a significant mass fractionation beginning at point A (Figure 3.5.b). This fractionation process moves the oxygen isotope datas along the dashed line (from A to B, C). It shows a slope of $\sim 1/2$. Now, the isotope values of the FUN's will be produced by mixing with "normal" matter (point D). In this connection, one of the most striking questions is whether the FUN's represent unaltered presolar matter in solid planetary material (f.e. meteorites) or whether they are the most in the present day Solar System primitive grains. In any case, the FUN inclusions are suitable candidates, because their oxygen isotope anomaly is correlated with other isotopic anomalies. A great number of other inclusion types has been examined to find such correlations. But, these results remained negative up today. However, there are some exceptions. The elements Ca, Ba, Nd, Sm, and Sr show also isotopic anomalies, whose abundances are established by other nucleosynthetic processes as compared with the ^{16}O production mechanism. Therefore BIBRING and LANGEVIN (1982) conclude a contamination of the solar nebula by different presolar sources. Further it has to be taken into account that homogenization took place because all investigated minerals show the same anomalies with the same magnitude also. In this way, an isotopic exchange could have caused a destruction of the primordial isotope pattern, at least partly. If that is a real picture it will be unlikely to find unaltered presolar grains in present day solar system solids.

CLAYTON et al. (1983) drew another picture on the base of oxygen isotopic investigations of chondrules of the Allende meteorite and ordinary chondrites. The results are shown in Figure 3.6.

Two isotope mixing events are implied in the Allende meteorite. In a "first run" the "Allende mixing line" is produced. A "second run" formed the isotope mixture of the chondrules from the first produced material. CLAYTON

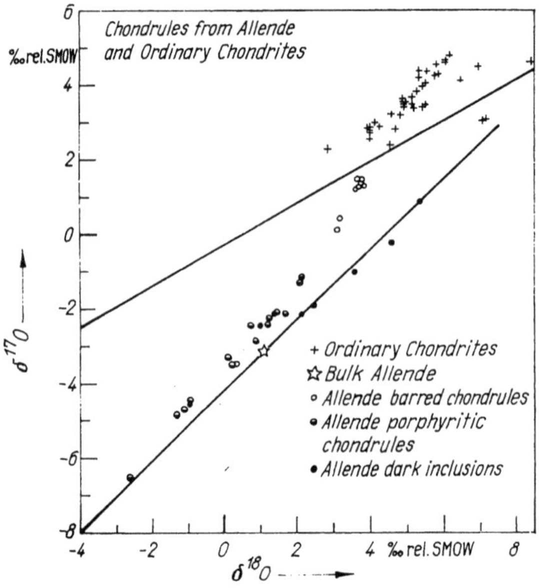

Fig. 3.6. Compositions of oxygen isotopes of chondrules from Allende and several ordinary chondrites (Allegan, Bishunpur, Bjurböle, Borkut, Cape Girardean, Chainpur, Chevra Hez, Dhajala, Hallingeberg, Krymka, Semarkona, Tieschitz). The "Allende mixing line" represents the composition of both white inclusions and dark inclusions (solid circles). Allende olivine-pyroxene chondrules trend towards the compositions of ordinary chondrite chondrules (CLAYTON et al. 1983).

et al. (1983) believe that the first formed "Allende mixing line" is a result of an exchange of ^{16}O-rich solids with an ^{16}O-poor gas. After the formation the chondrules were affected by a gas with another isotopic composition. The same gas contacted also the chondrules of the ordinary chondrites. This second gas was characterized by a similar isotopic pattern which controlled the isotopic compositions of achondrites, Moon and Earth. Now, CLAYTON et al. (1983) point out that if there were two gases with different compositions an inhomogeneous gas phase is not required necessarily. Inhomogeneities are possible to evolve as a result of reactions between gas and solids.

CLAYTON (1981) used two different grain populations in an other model. The carbonaceous chondrites originated from the first reservoir enriched in ^{16}O. The ordinary chondrites were formed from a second, ^{16}O-poorer source material.

Doubts arose recently concerning the propriety of the ^{16}O anomaly interpretations. THIEMENS and HEIDENREICH (1983a, b) discovered by means of laboratory experiments a mass-independent effect in the chemical production of O_3 from molecular O_2. The ozone was generated by electrical discharge in molecular oxygen, and a massindependent fractionation should have occured at the dissociation of molecular oxygen as a result of optical shielding by the major isotopic species ^{16}O, which is more abundant than ^{18}O and ^{17}O. But, THIEMENS and HEIDENREICH (1983a) are not able to state unambiguously whether shielding occurs on the electrons from the tesla coil or from ultraviolet radiation generated inside the reaction chamber of the experiment.

They found a slope of ~ 1 for the $^{17}O/^{16}O$ versus $^{18}O/^{16}O$ variation. AUDOUZE (1983) interpreted these results as follows. If the Solar System has been formed in an OB star association there was also a suitable UV flux to dissociate O_2. Therefore the ^{16}O anomaly could be explained in terms of the photochemistry of oxygen.

Magnesium isotope anomaly:

Considering the present knowledge about the Mg isotopic pattern of meteoritic constituents, the ^{26}Mg excess is produced apparently by an insitu decay of short-lived species. These ^{26}Mg excesses have been found in several inclusions of the Allende, Leoville, Murchison and related carbonaceous chondrites (LEE et al. 1976). LORIN and CHRISTOPHE MICHEL-LEVY (1978) and HUTCHESON et al. (1978) have shown that single crystals or mineral phases which were separated from inclusions exhibit a ^{26}Mg excess if the investigated material was Mg-poor and Al-rich. Further the ^{26}Mg excess is proportional to the ^{27}Al content (Figure 3.7).

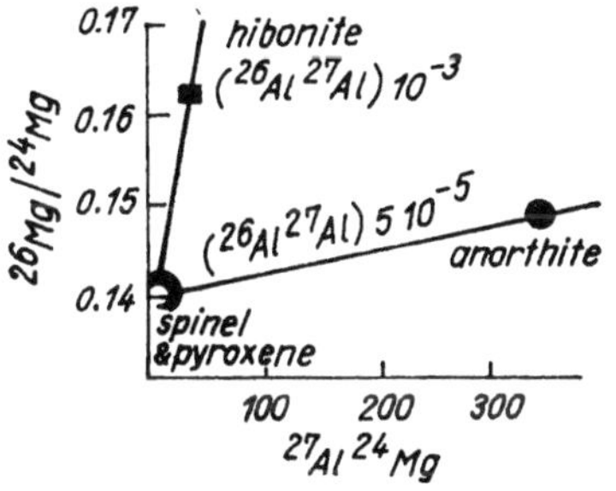

Fig. 3.7. $^{26}Mg/^{24}Mg$ versus $^{27}Al/^{24}Mg$ in mineral phases of related chondrite inclusions (BIBRING and LANGEVIN 1982).

The ^{26}Mg excess reaches up to 40%. These results allow the following conclusions:

1. The ^{26}Mg excess results from the decay of ^{26}Al ($T_{1/2} = 7.2 \cdot 10^5$ years). ^{26}Al was present in inclusions during the period of the crystallization.
2. After BIBRING and LANGEVIN (1982), the ^{26}Mg-rich constituents are uniformly distributed within all mineral phases which are contained in the inclusions. In comparison to that the ^{16}O excess was found only in few mineral individuals. The authors emphazise also that just the Mg-rich and Al-poor grains (without ^{26}Mg excess) are characterized by a composition which corresponds perhaps with that of the ^{16}O-rich phases.
3. CLAYTON (1977) pointed out that the ^{26}Mg-rich grains are presolar condensation products which were formed outside the Solar System.

Taking into account all possible processes which are able to produce these detected high ^{26}Mg-values, BIBRING and LANGEVIN (1982) pointed out that the most likely origin of the ^{26}Mg excess is an incorporation of fresh ^{26}Al by a supernova. On the other hand, the ^{26}Mg-rich grains could be also high temperature condensates of the solar nebula itself.

However, the time interval must be very small between injection of ^{26}Al and the end of the condensation and crystallization of the ^{26}Mg excess bearing mineral phases (either as a product of a supernova or by condensation in the solar nebula).

Moreover, the same time relation must be assumed for the nucleosynthetic event. All together, such "time tables" are also suitable for a largely inhomogeneous solar nebula.

The hibonite correlation line $^{26}Mg/^{24}Mg$ versus $^{27}Al/^{24}Mg$ (Figure 3.7) would suppose a very high initial $^{26}Al/^{27}Al$ ratio assuming a ^{26}Mg production by decay of ^{26}Al. After BIBRING and LANGEVIN (1982) this $^{26}Al/^{27}Al$ ratio must be $\sim 10^{-3}$. But, calculations have shown that such a value is implausible, because a ratio of $5 \cdot 10^{-5}$ could be rather real in the Solar System.

Neon isotope anomaly:

There are three Ne-isotopes which are stable. The isotopic composition is plotted in Figure 3.8.

As is shown by BOCHSLER (1983), there are the following Ne-isotope ratios:

$$^{20}Ne/^{22}Ne \approx 1;\ \text{spallogenic},$$
$$\approx 13.7;\ \text{solar wind},$$
$$\approx 9.8;\ \text{terrestrial},$$

$$^{22}Ne/^{21}Ne \approx 31 \pm 4;\ \text{solar wind},$$
$$\approx 34.5;\ \text{terrestrial atmospheric}.$$

Further, terrestrial neon is comparable with "planetary" neon as it was detected in carbonaceous chondritic meteorites.

A mixture of the three neon isotopes (spallogenic, solar wind, planetary) is shown in Figure 3.8. The pure components are identical with the corner points of the triangle. Now, it has been demonstrated by results of a stepwise heating of the Orqueil meteorite that the extracted neon did not lie within this "normal" triangle between 800 °C and 1100 °C (JEFFERY and ANDERS 1970). BLACK (1972) interpreted this effect in that sense that there is a ^{22}Ne-rich component, the so-called Ne—E, which could be produced by nucleosynthetic events outside the Solar System. On the base of similar experiments, by EBERHARDT et al. (1981) have been concluded that ^{22}Ne was produced by decay of ^{22}Na. These authors advocated this opinion, because they found the ^{22}Ne to a very high degree of purity. Other formation processes (nucleosynthesis or irradiation) are not able apparently to yield such ^{22}Ne. If ^{22}Na is the source of ^{22}Ne the incorporation of ^{22}Na in the solid grains must have happened very rapidly, because the ^{22}Na half life time is only 2.6 years. Contrary to this,

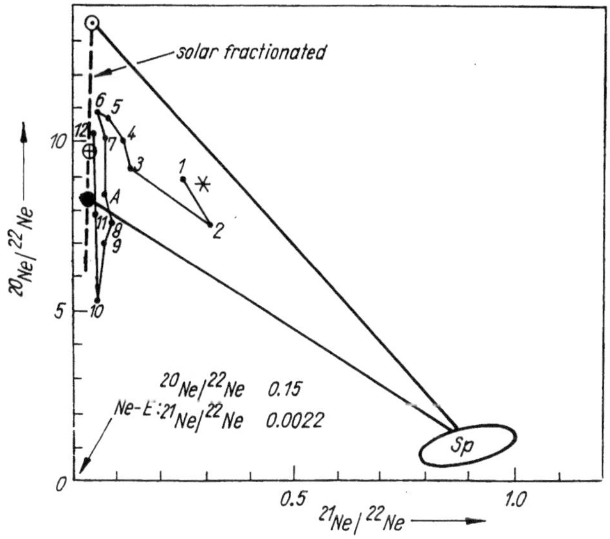

Fig. 3.8. The isotopic composition of neon (BOCHSLER 1983).
The isotope types are:

⊙ — solar wind
⊕ — terrestrial atmospheric
sp — spallogenic
A — planetary
X — analyzed neon by stepwise heating from the Orgueil meteorite, numbers represent temperature steps in 100 °C.

^{22}Na could be produced also by energetic irradiation of elements like Mg, Al, Si or F in the grains. But, this irradiation must have been stopped before the carriers lost the Ne. Otherwise, ^{21}Ne and ^{20}Ne ought to be detected noticeably (BOCHSLER 1983).

Isotope anomalies of Calcium, Silver, Mercury:

LEE et al. (1978) reported about excessive ^{48}Ca found in an exotic inclusion of the Allende meteorite. These results were confirmed by NIEDERER and PAPANASTASSIOU (1979). They investigated a black chondrule (Ca, Al-rich) and got a positive excess of 5‰.

KELLY and WASSERBURG (1978) observed excessive ^{107}Ag in the Santa Clara iron meteorite. In this connection, LAVRUKHINA (1980) pointed out that resulting investigations of both the $^{107}Ag/^{109}Ag$ and $^{110}Pd/^{109}Ag$ ratios and the slop of the Pd/Ag-isochron gave an evidence for the in situ decay of ^{107}Pd (half life time = $6.5 \cdot 10^6$ years).

YANAGITA and GENSHO (1977) investigated the $^{202}Hg/^{196}Hg$ ratio of some carbonaceous chondrites (Orgueil, Murchison, Allende) and several ordinary chondrites (Tieschitz H_3, Allegan H5). They found an occasional increase of the $^{202}Hg/^{196}Hg$ ratio, what might be caused by decay of ^{202}Pb (half life time $= 3 \cdot 10^5$ years).

Xenon isotope anomaly:

Xenon shows large variations in various meteorites in contrast to neon and other noble gases. According to LAVRUKHINA (1980) the isotope composition of terrestrial, lunar, and iron meteorite specimens are almost identical, whereas carbonaceous chondrites and ureilites show an other xenon isotope pattern. Further, the solar wind xenon differs from all known sites (meteorites, Moon, Earth).

Considering only the carbonaceous meteorites, there is a simultaneous enrichment of both the light and the heavy xenon isotopes. The main carriers of these isotopes are black residues left after treating the meteorite samples with non oxidizing acids (LEWIS et al. 1975). These residues consist of amorphous carbon, chromite, spinel, and a widely unidentified phase called "Q"-phase. After LAVRUKHINA (1980) the "Q"-phase contains up to 90% of all Ar, Kr, and Xe. The "Q"-phase was identified especially in the Allende meteorite. However, the unetched residues of the Murchison and Orgueil meteorites contain similar phases and additionally two unidentified carbonaceous phases (C_α, C_β).

The heavy xenon isotopes are enriched in chromite in all carbonaceous chondrites. ^{134}Xe and ^{136}Xe are concentrated in Allende meteorite also in the carbon phase. The light xenon isotopes are enriched in both the carbon phase and chromite. A ^{130}Xe excess was found in the so called C_β-phase of the carbonaceous chondrites Murchison and Orgueil. Further the Allende spinel phase contains an excess of such "medium" Xe isotopes, but it is depleted in light and heavy isotopes.

Planetogonic conclusion

Doubtless, the isotopic anomalies have played recently a significant role at the examination of questions about the origin and evolution of the planetary system. Until few years ago it was accepted generally that the primordial solar nebula was a well mixed system of matter. High temperatures prevailed and therefore there were no solids to begin. But if solids occured, they all were vaporized again and the remaining constituents admixed into the "rest" of the nebula.

The present day picture is more complicated, because we know the described isotopic anomalies among other things.

Apparently dust has played a substantial role, especially in connection with the above mentioned isotopic anomalies.

Isotopic anomalies must be investigated in such matter that has not been altered since its formation. However, today it seems to be that all material available for such analysis is metamorphosed. The FUN-inclusions might be the only constituents which show the original isotopic composition. They were observed but only in three inclusions of the Allende meteorite (up today).

One of the most significant questions is, whether the anomalous isotopic compositions are extraneous, or wheather it has been produced within and during the evolution of the Solar System. Begemann (1980) pointed out, that if the anomalies were produced within the Solar System only parts of the nebula material must have been affected or the production of diverse isotopic patterns was effective not in the same kind in the whole system. A suitable initial process could be the irradiation with energetic particles. (see sect. 3.4.5.). The radioactive nuclides which were produced on this way could have changed the original isotopic abundances by their decay as well as by direct transformations. After Begemann (1980), the difficulty is not to find a source which produces sufficient protons but an irradiation mechanism which allows a rather special energy spectrum. In this connection, Lee (1979) estimated proton fluxes of 10^{18} cm^{-2}, 10^{21} cm^{-2}, and 10^{25} cm^{-2} to produce the abundances of Ne—E, ^{26}Al, and ^{16}O, respectively (more details in sect. 3.4.5.).

To irradiate the whole solar system in such a "special" manner would surely not be necessary. Begemann (1980) points out that a formation of these isotopic abundances only in narrow regions of the solar nebula and a later admixture to unexposed remainders would have the same effect.

This scenario is able to explain a process which yields the isotopic composition for an individual element. But, known isotopic anomalies in general can not be produced by this process alone. To enrich Neutron-rich isotopes of the heavier elements is still more difficult as well (Begemann 1980).

The isotopic anomalies could represent unaltered remnants of stellar processes, as mentioned above. This opinion agrees with present days ideas that the abundances of the elements are caused by a mixture of components which were formed by several nucleosynthetic processes and also at different times and places. Taking into account the present day knowledge the following scenario is conceivable:

The presolar material cloud consists of gaseous matter as well as solid grains. The solid grains are primary condensates formed in expanding novae, supernovae, etc. (Fig. 3.1). Reeves (1978) formulates that the Solar System was "born admidst a fireworks of supernovae". That means, the proto-solar nebula was part of a huge great material cloud. The presolar nebula was characterized by a standard chemical composition with about 1% dust. The origin of the Solar System began with the condensation of gaseous components (eventually triggered by injection of solids) and the injection of matter synthesized in exploding stars. These processes, including nucleation and the first accretion also took place very fast. The Ne—E abundances and the ^{26}Mg excess support this conception, because the half life times of ^{22}Na (Ne—E) and ^{26}Al (^{26}Mg) are only 2.6 years and $0.76 \cdot 10^6$ years, provided that Ne—E (^{22}Ne) and ^{26}Mg were formed by the decay of ^{22}Na and ^{26}Al respectively.

As mentioned it is conceivable that the condensation and also the collapse of the solar nebula could have been triggered by a supernova. The expanding shock front caused the first events and as by BIBRINHG and LANGEVIN (1982) suggested this supernova matter was rich in ^{16}O. On this base these authors discuss the different ^{16}O enrichment in the investigated inclusions of meteorites. The supernova shell is progressively diluted by the solar nebula. Depending on elapsed time and increasing distance from the supernova the least refractory grains contain the least ^{16}O as observable in some inclusions. This result was confirmed by WARK (1979). He showed further that the greatest and most coarse-grained inclusions exhibit the most significant ^{16}O excesses. Therefore, the coarse-grained inclusions were formed closest to the supernova. They originated at first. With sinking temperature this supernova material was "consumed" more and more. Now, condensates arise which were formed at deeper temperatures. This happened in an essentially "pure" presolar nebula. The matrix material of carbonaceous chondrites (CI-type) consists of such "late" condensates. They contain the above mentioned ^{22}Ne also. Therefore, BIBRING and LANGEVIN (1982) suggest that the ^{22}Ne containing solids in carbonaceous chondrites are rather presolar than interstellar. Furthermore, these authors advocate that the ^{22}Ne enriched solids arose from quick-evolved and short-lived stars in the vicinity of the solar nebula. Contrary to this, the ^{16}O-rich grains must result from a supernova injection. CAMERON and TRURAN (1977) and LATTIMER et al. (1978) support also that star and/or supernova injections are responsible for the detected isotopic anomalies and that a shock front could has been the trigger mechanism for the collapse of the solar nebula.

All together, the isotopic anomalies point to both an imperfectly mixed preplanetary material reservoir and condensation, nucleation, and "first step" accretion processes which operated very quick.

3.4.5. Influence of Protosolar Activity (H. SPONHOLZ)

The T Tauri Stage of the Sun

During the last years increasing evidence has confirmed that there are general similarities between present T Tauri stars and what we assume about the early solar system. In addition to the arguments coming out of models for stellar evolution, the belief that the early Sun passed through an active, luminous T Tauri phase has for many years been based on the necessity to have a physical cause for the removal of excess mass by sweeping outward gas and dust left over by planetary formation. Such a necessary mechanism may be the interaction between the ancient Sun surrounding matter and the solar wind expelling during the T Tauri phase at a rate much higher than observed today. The higher fluxes of photons and particles originating from such T Tauri state may have had important evolutionary consequences if they had occured in tho oarly Sun.

The possible passed T Tauri stage of the Sun may be also interesting in connection with the supernovae trigger hypothesis for the origin of our solar system. CAMERON and TRURAN (1977) made a number of suggestions chiefly connected with the discovery of excess ^{26}Mg in inclusions in the Allende meteorite which presumably formed as ^{26}Al whose half-life is about 0.7 m.y. They concluded that thus the inclusions must have formed within about 1 b.y. of the "last supernova" and suggested a picture in which the formation of our protostar was triggered by some nearby supernova within a billion-year time scale.

FOWLER et al. (1962) discussed at first the possibility that short-lived radionuclides e.g. ^{26}Al, ^{107}Pd, and ^{129}I might be produced by irradiation within the solar system. HEYMANN and DZICZKANIEC (1976) have outlined the occurance of ^{26}Mg anomalies in terms of proton bombardment of a gas of solar composition. CLAYTON et al. (1977) calculated a number of nuclear cross-sections important for the idea that energetic-proton irradiation of grains during the formation of the solar system may be responsible for several isotopic anomalies discovered in meteorites. According to a study of WORDEN et al. (1981), based on observations of short period photometric fluctuations in T Tauri stars, explicable as a superposition of many solar-like flare events, one may explain isotopic abundance anomalies without recourse to nearby supernovae. FEIGELSON (1982) discussed a more accurate measure of activity in solar-type stars with the help of their soft X-ray luminosity, obtained from the observations with the Einstein X-ray Observatory.

Flare Activities of T Tauri Stars

There are relevant observational results supporting a general trend of higher X-ray luminosities in younger stars. On the basis of studies of chromospheric emission SKUMANICH (1972) attemped to quantify the correlation of stellar activity with stellar age as a scaling law in which activity varies as the inverse square root of the age. Studies of both coronal X-rays (STERN 1984) and the connection of X-ray luminosities with age of the Orion population ($\sim$ 3 m.y.), the Pleiades ($\sim$ 60 m.y.), and the Hyades ($\sim$ 600 m.y.; CAILLAULT and HELFAND 1985, MICELA et al. 1985), tend to confirm this decrease in stellar activity with age (for the time after 1 b.y. CAILLAULT and HELFAND argue rather for a rapid exponential decline), although the exact functionential relationship remains unclear.

Several studies performed with the Einstein X-Ray Observatory launched in 1978 (High Energy Astronomical Observatory HEAO-2 satellite) have shown that many early solar-type stars have X-ray luminosities much higher than the present Sun's emission. The Einstein satellite X-ray telescope was first used to resolve a number of discrete X-ray sources in the central region of the Orion nebulae by KU and CHANAN (1979). Fifty-eight discrete sources were found, most of them may be pre-main-sequence stars, but only 5 at flux levels of $3 \times 10^{30} - 5 \times 10^{31}$ erg/s are identified as pre-main-sequence stars of the Orion Population according HERBIG and RAO's (1972) catalog (see COHEN 1984). For the Taurus-Auriga association 14 T Tauri stars were detected, and their luminosity in the 0.1–4.5 keV range is $5 \times 10^{29} < L_x < 8 \times 10^{30}$ erg/s

(Feigelson and de Campli 1981, Walter and Kuhi 1981, 1984). Montmerle et al. (1983) found in the Rho Ophiuchi Dark Cloud forty-seven X-ray sources with luminosities ranging from 5×10^{29} erg/s to 10^{32} erg/s including 9 of the 11 previously known T Tauri stars contained in the investigated area.

An important characteristic of the X-ray emission associated with T Tauri stars is the rapid and large amplitude luminosity variations, which strongly points toward for the existence of "superflares" on pre-main-sequence stars. The flux from the T Tauri star DG Tau increased by an order of magnitude in only 10 min with a peak luminosity of 4×10^{31} erg/s (Feigelson and De Campli 1981). Walter and Kuhi (1984) reported a large, slow flare in AS 205 (a source in the Taurus Auriga cloud complex) with a rise by more than a factor of 10 in X-ray flux to a maximal value of 1.5×10^{31} erg/s within 5×10^{3} s, followed by a decrease to the quiescent level over 10^{4} s. The Rho Ophiuchi T Tau candidate star ROX 20 is quite remarkable. It is one of the most energetic of all the X-ray flare events yet seen. Montmerle et al. (1983) observed a maximum of X-ray luminosity of 10^{32} erg/s decreasing within a day to a quiescent level of 8×10^{29} erg/s. These three variable sources are not peculiar objects. The observations of the Oph Rho region span some 2 years, over which interval almost all the sources vary with the following qualities: The average peak-luminosity runs to 5×10^{30} erg/s, one flare out of two has a lumonisity twice that of the quiescent level, and these flares contain half of the total X-ray intensity measured in the cloud (Montmerle et al. 1983). Five of the 14 observed pre-main-sequence stars in the Taurus Auriga cloud complex showed significant X-ray variability on long time scales of about years. This should not be surprising — the variability may be due to changes of large-scale structures of active regions as is seen in the solar cycle (Walter and Kuhi 1984). On the other hand, the short-time scale X-ray luminosity variations of pre-main-sequence stars reported above are similar to that observed in solar flares, and one may assume that solar-like flaring may account for that variability (Montmerle et al. 1983).

The main distinction between the solar and the T Tauri X-ray emission seems to be the order of magnitude observed. For the contemporary Sun most authors adopt an average soft X-ray luminosity of 3×10^{27} erg/s, that is more than three orders of magnitude below that seen in T Tauri stars. From the solar-flare energy, T Tauri flare energy, and from the average solar-flare proton flux one can estimate the T Tauri proton flux. The energy spectrum of solar protons is soft and can be represented fairly well as a power law according to $E^{-\gamma}$, where E is the kinetic energy per proton, and the values for γ ranging from 2 to 4 for energies between 20 and 80 MeV and are lower for lower energies (Reedy et al. 1983). The solar proton flux varies strongly with the 11-year cycle during which a few large flares produce most of the solar protons. During the largest flares in the contemporary sun, the proton flux with particle energies more than 6 MeV has been reported to be $\sim 10^{3}\,\text{cm}^{-2}\,\text{s}^{-1}\,\text{ster}^{-1}$ (Feigelson 1982, see also Vahia and Biswas 1982).

Assuming that proton and X-ray fluxes scale linearly one can expect proton intensities in T Tauri stars of $10^{6}-10^{7}\,\text{cm}^{-2}\,\text{s}^{-1}\,\text{ster}^{-1}$ appropriate to the observed X-ray flares in pre-main-sequence stars. Worden et al. (1981)

estimated a proton flux of $2 \times 10^7 - 2 \times 10^8$ cm^{-2} s^{-1} ster^{-1} based on the optical variations in T Tauri stars. However HUDSON (1978) presents evidence that a flare's peak proton flux would vary as the 5.6 power of its flare energy. With the help of this relation one obtains proton fluxes more than ten orders of magnitude stronger than estimated above. Although it seems that Hudson's relationship is not applicable to such extreme range, one can assume that the dependence of proton flux on the power of X-ray flares is possibly a nonlinear function.

Isotopic Anomalies and the Activity of the Early Sun

The protons in the flare induced cosmic rays from the early sun have sufficient energy to interact with preplanetary matter in the ancient solar system and to generate reaction products that persist for long periods of time. These reaction products can be used to study the history both of the solar activity and the bombarded targets such as the moon and the meteorites. The two main interactions of energetic nuclear particles are 1) nuclear scattering and 2) nuclear reactions. The presently available lunar samples represent different epochs during the last 1 b.y. In coincidence with the astrophysical stellar estimations reported above they indicate that during the last 10 m.y. the average fluxes of solar protons are similar to those estimated recently, although there have been very few solar flares that are orders of magnitude larger than those observed during the last two decades (compare GOSWAMI et al. 1984). The observed solar flare records in some meteorites refer to very ancient times especially in the case of carbonaceous chondrites. The time of irradiation of the grains range between 4.2 b.y. and 4.45 b.y. before present with a total irradiation time less then 1 m.y.

Numerous nuclear isotopic anomalies in meteorites were discovered during the last decades (see e.g. BEGEMANN 1980, WASSERBURG et al. 1980). Models about the origin of some of these anomalies require either injection of material from nearby stellar thermonuclear reactions or irradiation by solar proton flux several orders of magnitude higher than today.

The subject of many searches have been magnesium isotopic anomalies, strictly speaking an anomaly characterised by an excess of ^{26}Mg, which have been found in several inclusions with high Al/Mg ratios (see sec. 3.4.4.). GRAY and COMPSTON (1974), LEE and PAPANASTASSIOU (1974), and later LEE et al. (1076) discovered magnesium isotopic shifts of 0.2%–0.4% respectively 1.3% in certain Ca–Al rich inclusions from the Allende meteorite. LEE et al. suggested that the ^{26}Mg excess is due to the decay of now extinct ^{26}Al, and this assumption is now generally accepted. A possible process which could have given rise to ^{26}Mg is a proton irradiation of circumsolar material, which causes the reaction ^{26}Mg(p, n)^{26}Al and after chemical separation and decay, the ^{26}Mg anomaly. The average particle fluence needed to explain the ^{26}Al excess, was estimated by CLAYTON et al. (1977), (see also LEE 1979), as 1×10^7 cm^{-2} s^{-1} of $>$ 6 MeV protons for 1×10^6 years duration, a value which is consistent with that expected from T Tauri stars (FEIGELSON 1982).

Similarly, HEYMANN and DZICZKANIEC (1976) discussed an analogous reaction on ^{22}Ne produces ^{22}Na. In situ decay of the ^{22}Na, afterwards mass-dependent fractionation would lead to an enrichment of the ^{22}Ne component as observed in several carbonaceous chondrites e.g. the meteorite Orgueil (EBERHARDT et al. 1981). However, CLAYTON et al. (1977) came to the conclusion that such an origin of the ^{22}Ne-anomaly is highly unlikely, because the half-life of ^{22}Na (2,6 yr) in comparison with the time scale, in which diffusion or chemical separation occurs, is just too short, because of the relative high volatility of the produced noble gas and the high temperatures of the irradiated grains. An alternative process, in which ^{22}Na is produced by proton irradiation of ^{25}Mg [according to $^{25}Mg(p, \alpha)^{22}Na$] appears improbable because the suspected chondrites with ^{22}Ne-anomaly are apparently Mg-poor.

A well-identified short-lived nuclid is the extinct ^{129}I whose half-life is 16 m.y. The general accepted evidence for the existence of ^{129}I in the ancient solar system is the presence of ^{129}Xe anomaly correlated with the presence of ^{127}I, which was shown for meteorite samples by JEFFREY and REYNOLDS (1961). This correlation has been found in many samples; ^{129}I must have been widely distributed throughout all early solar nebulae condensates with an isotopic ratio of $^{129}I/^{127}I \approx 10^{-4}$. CLAYTON et al. (1977) considered an origin of the short-lived ^{129}I caused by proton irradiation of an early active Sun. For an assumed total proton flux of 2.8×10^{20} cm^{-2} they found isotopic compositions equal to the mean compositions observed in meteorites. This is in agreement with the proton fluxes estimated above.

CAFFEE et al. (1983) analysed individual irradiated grains (showing solar flare particle tracks) and unirradiated grains (that do not show such tracks) from the Murchison CM chondrite and the Kapoeta horwardite. They found that irradiated grains from Murchison meteorite contain 30 times more cosmogenic ^{21}Ne than their unirradiated counterparts and estimated from the anomaly, the measured target element abundances, and the relevant production rates from Al, Mg, Si, and Fe a duration of the exposure of about 10 m.y. for the irradiated Murchison grains, and about 20 m.y. for Kapoeta, based upon the current average solar proton flux at a heliocentric distance of 1 AU. The precompaction exposure duration, estimated from solar particle-track densities, is comparatively two to four orders of magnitude shorter. A possibly enhanced solar flare activity as the sun evolved through a T Tauri phase may explain this discrepancy. PADIA and RAO (1985) studied pyroxene mineral of the Kapoeta meteorite and obtained from the abundance of solar cosmic ray spallation neon, proton fluxes from the early sun of about 3 to 4 orders of magnitude above than those measured today.

The possibility that the isotopic anomaly in oxygen has been generated by a local irradiation mechanism in the early solar system has been addressed in several papers (HEYMANN and DZICZKANIEC 1976, CLAYTON et al. 1977). The authors attempt to explain the ^{16}O isotopic anomaly as destruction of the ^{17}O and ^{18}O in the target by (p, α) reactions so that only ^{16}O survived. It should be noted that the oxygen anomaly is not easily produced by means of a particle irradiation process of the type discussed in connection with the other anomalies. The main difficulty is the high abundance of oxygen, which necessitates such a high particle flux that one would expect to observe solar flare induced anomalies in a number of elements, and evidences of such effects are not present.

The composition of dust particles from the comet Halley has been analyzed by the mass spectrometer (PUMA) of the spacecraft VEGA. Isotopic ratios have been measured for Carbon, Magnesium, Silicon or Iron; they are rather close to the normal solar system values (Langevin et al. 1987, Jessberger and Kissel, 1987, Wäsch, 1986).

Isotopic Anomalies — Products of Irradiation or Extraneous Origin?

There are still a number of uncertainties in the interpretation of isotopic anomalies. Several models are invoked to explain some or all of the anomalies. However, each of these explanations have been more or less subject to criticism. The suggestion that the formation of the solar system had been triggered by a supernova remnant was based mainly on the evidence for the presence of short-lived isotopes in the early solar system, especially of the short-lived nuclei ^{26}Al (Cameron and Truran 1977, Clayton 1979). In the framework of this theory the supernova may have been responsible for injecting of short-lived nuclei into the presolar cloud. However, it has been remarked by W. A. Fowler, that "^{26}Al could not be synthesized in supernovae at high temperatures, where neutrons are copiously produced because of the expectation of a large cross section for $^{26}Al(n, p)^{26}Mg$" (Fowler 1985). A second argument against this theory for the origin of ^{26}Al, is the recent detection of interstellar ^{26}Al. The HEAO 3 spacecraft, which has searched for diffuse gamma-ray emission from the galactic equatorial plane, recorded a discrete line in the spectrum at 1809 keV, coming from the general direction of the centre of our Galaxis. This peak is a well known radiation result of the decay of the ^{26}Al by $^{26}Al(e^+\nu)^{26}Mg^*(\gamma)^{26}Mg$ (Mahoney et al. 1984). Possible origins of the ^{26}Al in the interstellar medium are nova explosions (Clayton 1984), or gas streaming out from giant stars. The estimated isotopic ratio of ^{26}Al to ^{27}Al, that is 10^{-5} in the interstellar gas, is rather close to that provided by meteorite samples, suggesting that there is no need for assuming a supernova trigger hypothesis (Clayton 1985). Clayton and Leising (1987) argued that supernova nucleosynthesis cannot produce an isotopic ratio of the interstellar medium much in excess of $^{26}Al/^{27}Al = 1 \times 10^{-7}$. Cameron (1984) recently modified his hypothesis and suggested that red giant stars can account for the presence of ^{26}Al in the presolar matter. Another possible source of the radioactive aluminium, are Wolf-Rayet stars (Dearborn and Blake 1985). However, the enrichment of ^{26}Al in some meteorite samples is too large to be interpreted as being the average interstellar abundance. Thus, there must have been some additional source of ^{26}Al enhancement in the presolar cloud.

A production mechanism of the short-lived nuclides ^{26}Al, ^{129}I, ^{107}Pd, and other isotopic anomalies, could be also the result of an intense irradiation of protoplanetary material by solar flare protons during the T Tauri phase of the Sun. This requires that the gas be early dissipated and that the proton bombardment acts on the solid components, due to the strong ionisation loss in gas-rich matter (see WASSERBURG 1985). The weakness of the irradiation model lies in the necessity to have a rather special energy spectrum of the protons, and on the different orders of proton fluxes needed to explain the several anomalies. For example, the proton fluxes needed to produce the observed anomalies of ^{22}Ne, ^{26}Mg, and the abnormal oxygen isotopes are approximately 10^{18} cm^{-2}, 10^{21} cm^{-2}, and 10^{25} cm^{-2}, respectively. A clear indication against a local irradiation mechanism for ^{26}Al as the only cause for the ^{26}Al-anomaly, is the recent discovery of the existence of ^{26}Al in the interstellar medium today. This ^{26}Al certainly cannot have been produced in the way of solar irradiation (FOWLER 1985). It is also difficult to obtain by irradiation the enrichments of the neutron-rich isotopes as observed in some isotopic patterns of meteorites. Although the variation of isotopic ratios in meteorites due to the influence of solar cosmic protons can be calculated on the basis of known cross sections, the properties of the secondary nucleon fields and their contribution to the production of cosmogenic nuclides remains an unresolved question and needs further experimental efforts (see R. MICHEL et al. 1986).

In fact, it is not unlikely that none of the discussed production mechanism alone is responsible for all anomalies, that the influence of both irradiation and the extraneous sources caused the isotopic anomalies.

3.4.6. Mineralogical-petrological and Physical Implications

The most suitable information about the earliest stages and chronology of the solar system yield chemical datas and the radioactive dating of planetary material. Mineralogical-petrological and physical propertie f.e. do not have a comparable conclusiveness. Above all, this is because there are no appropriate investigations. Another reason is that suitable planetary material, allowing laborative examinations, is hardly available scarely.

Nevertheless, mineralogical-petrological and physical investigations could contribute to a better understanding in any case.

Mineralogical implications:

The significance of unaltered preplanetary material for the investigation of "planetary" questions is quite well known. In this connection, one of the most suitable candidates are carbonaceous chondrites and some constituents of this meteorites, especially the refractory inclusions, generally known as Ca—Al-rich inclusions (CAI's) in type II and Type III carbonaceous meteorites. Their bulk compositions are dominated by high enrichments of refractory elements (factor 15—10^4) compared to bulk carbonaceous chondrites. The main components are Al_2O_3, CaO, MgO, SiO_2, and TiO_2. Volatile elements are less abundant. Above all a mineral phases occur melilite, spinel, and perovskite (type A inclusions) and melilite, fassaite, spinel, anorthite, and perovskite (type B inclusions). Some type A inclusions contain a crystalline Al_2O_3 phase and hibonite ($CaAl_{12}O_{19}$). Compounds like $CaAl_4O_7$ and $CaAl_2O_4$ are also possible. $CaAl_4O_7$ has been proven at first in an inclusion of the Leoville carbonaceous chondrite (Christophe Michel-Levy et al. 1982). These authors investigated Leoville CAI's and they found in a coarse-grained inclusion a granular intergrowth of gehlenite, spinel, perovskite, hibonite and $CaAl_4O_7$. The bulk composition is clearly dominated by gehlenite. Apparently, $CaAl_4O_7$ belongs to the earliest condensates, what results from calculations of a typical condensation sequence of a cooling solar nebula. Corundum (Al_2O_3) and hibonite ($CaAl_{12}O_{19}$) follow in this condensation sequence. Contrary to this Christophe Michel-Levy et al. (1982) point out or think that these phases rather point to an igneous origin, because of the existing mineral association and contact relationships in this association. Probably, there was a slightly metamorphic event also. The $CaAl_4O_7$ phase is probably stable at high temperatures. Therefore, the presence of $CaAl_4O_7$ could be an indication of a high temperature history of the investigated CAI's. On the other hand, the absence of $CaAl_4O_7$ in the most other CAI's suggest metamorphism at lower temperatures. A further argument against a high temperature condensation origin is the appreciable amount of Fe and Cr in the mineral phases. Fe and Cr are moderately volatile elements. They should not be present in high temperature condensates. Moreover, these elements must be of primary origin (f.e. because of the distribution between individual phases, crystallization history, diffusion properties). All together, the authors propose that the investigated inclusion represents a residue of incomplete volatilization., The presence of moderately volatile elements (Fe, Cr) is explainable only, because the final abundances of volatile elements are governed by the complex melt-vapour distribution coefficients. Contrary to this, a gas-solid nucleation is a very effective separation process for elements with different volatilities.

Finally, Christophe Michel-Levy et al. (1982) found three rims around the investigated CAI. The innermost rim contained large amounts of hibonite and it is the most refractory one. This rim was formed apparently by loss of moderately volatile components from the inclusion surface. The other rims consist of diopside and olivine. They could have been developed by condensation. The condensation of the two minerals arose not from a gas of solar composition, but from a gas which was partially evaporated from preexisting rocks. This final conclusion emphasized, that in the investigated CAI there is "partial evaporation and partial recondensation" clearly documented.

Kurat et al. (1975), received similar results from investigations of a Ca—Al-rich chondrule of the Bali carbonaceous chondrite. The chondrule is strongly enriched in refractory elements. Further it was shown that refractory elements like Ti and Al are concentrated in or near the surface of the condrule. In contrast, volatile elements like Na, Si, and Cr are enriched in the innermost parts. Kurat et al. (1975) suggest that this chondrule was formed not by condensation from a nebula of solar composition. The chondrule originated rather "in one, or more likely in several impact events and that the element distribution is the result of volatilization-condensation".

The results, received by Christophe Michel-Levy et al. (1982) and Kurat et al. (1975), show very impressively that these meteoritic components do not represent first condensates of a solar nebula. They are solids which were formed by many complicated processes rather. A significant role was played by impacts, collisions between solid grains and greater planetary objects, volatilization, evaporation, and partial recondensation.

The results demonstrate also that it would be very difficult, perhaps even impossible, to find true "first" condensation products. The investigations show that the interaction between small solid particles was very intensive process energetically. On the other hand, many meteorites have never been affected noticeably by differentation events since their formation (type CI-chondrites). Because there are CAI's which are embeded in a "deep-temperature" matrix, gentle interactions ("soft" accretion, agglomeration) have also been possible. Thus, the matrix material of carbonaceous chondrites could be of more primitiveness than the inclusion (like CAI's). But, the knowledge of the matrix ins still very slight, because experimental investigations of matrix matter are more difficult than those of the comperatively well defined inclusions, chondrules, and mineral aggregates.

Hutchison and Bevan (1983) investigated the petrography and mineralogy of some H-group chondrites. They favour the formation of ordinary chondrites by the hot accretion of chondrules. Furthermore, no direct evidence of a solar nebula may exist. Other authors postulate the presence of a solar nebula during the accretion. Taking into consideration the results of Christophe Michel-Levy (1981) and other workers which argued that chondrules formed by the melting of pre-existing differentiated solids (that means differentiation only for metalsilicate fractionation) Hutchison and Bevan (1983) argue more extensive. They suggest fractionated planetary precursors for chondrules and their associated clasts and liquids. This model requires the existence of both, a volatile poor lunar or eucritic planetary body and a gaseous protoplanet before the accretion of condrites. This model bases on interactions (f.e. collisions) of precursors during the genesis of ordinary chondrites and chondrules also. The most significant conclusion for the origin of planetary bodies is that differentiation in planetary bodies had occured very early.

Opposite to this, Grossman and Wasson (1983), investigating chondrules of carbonaceous and ordinary chondrites, suggest that "some chondrule precursor materials must have reacted with nebula gases at temperatures $< 600\ °K$". They stated from variation of lithophile elements, content and ratio of volatile elements and/or refractory elements, etc., that the chondrule precursors could not have been pre-existing igneous or similar differentiated rocks. These solids result from condensation and reaction in the solar nebula.

Both of these interpretations do not contradict the model of a very fast condensation, nucleation, and accretion of planetary material to form planetary sized bodies.

Also by investigations of the crystallization history of chondrules the number of possible interpretations of their origin and of the formation conditions of preplanetary material have been reduced.

Dynamic crystallization studies suggest that chondrules were heated up to a range of 1400–1600 °C. The cooling rates ranging from a few degrees to some thousand degrees per hour down to a relatively low ambient temperature (HEWINS, 1983). Further, HEWIN's (1983) review indicates that his working hypothesis is consistent with the melting of primordial particles to generate chondrules. His conclusion may rule out a number of other kinds of chondrule origin. Especially also condensation and impact models could be set aside. At first, because for textural but also isotopic reasons.

WOOD (1983) proposed that chondrules were formed by the settling of dust particles toward the median plane of the solar nebula where they were heated by friction. An other process is the heating of precursors by flares. The great variety of chondrite constituents could possibly be explained on the base of both mechanisms. They might have been formed by many different mechanisms, depending on infalling distance of the precursors, interactions of droplets with solid particles, partial evaporation of melted material, f.e. to produce the CAI's (KING 1982).

The matrix of carbonaceous chondrites (above all the CI-types) consists of fine grained layer lattice silicates (serpentine, chlorite, montmorillonite). One of the most significant questions is now, whether such matrix material might has been present in the Solar System, especially before solar flare irradiation began. In addition to this, GOSWAMI and MACDOUGALL (1983) carried out nuclear track and compositional studies of meteoritic olivines. The authors found that the investigated single grains, grain aggregates, chondrules, and bulk olivine separates show evidence of solar flare irradiation (sect. 3.4.5.). But nonirradiated grains also were detected. Furthermore, the solids showed isotropic as well as anisotropic irradiation. The irradiation apparently took place in the near-surface region of a small object (up to 1 meter in size) under variable shielding conditions, according to the model of GOSWAMI and LAL (1979).

Some of the analyzed grains have quite complex irradiation patterns. This complexity could be originated from partial disaggregation of these small bodies with following mixing and formation of objects up to kilometers in size before or during agglomeration. Both, irradiated and nonirradiated components (olivines) are characterized by the same major and minor element composition. This suggests that olivine separates belong to a single population which existed before compaction and irradiation. The investigated samples seem to be a mechanical mixture.

Unfortunately, the experiments described above yielded no results concerning the matrix silicates. Nevertheless, it has been shown by MACDOUGALL and PHINNEY (1977) that the matrix silicates of the Murchison-CM-meteorite contain solar wind ions. The layer lattice silicates were possibly irradiated at the same time as the inclusions. This idea could be plausible, because the matrix aggregates would have provided appropriate shielding material (after MACDOUGALL and PHINNEY 1977). H_2O-ice would provide a similar effective shielding. The inclusions are embedded in an icy matrix during solar flare irradiation. This ice may also be the liquid H_2O-reservoir that is contained up to 20% in CI-type (CM) carbonaceous chondrites. If that is a real scenario, the CM-carbonaceous chondrites would have been formed nearby the asteroidal belt.

In principle, the presence of H_2O-ice could be the supposition to produce the layer lattice silicates of the chondrite matrix in general. If silicate condensates or clusters can contact H_2O-ice at suitable conditions, reactions between H_2O and silicatic matter should also be possible to form such layer lattice silicates. Supposing low temperatures and low pressures, an appropriated silicate might be serpentine. That could be the first originated "deep"-temperature matrix silicate, because serpentine will be disintegrated at first if the temperature rises.

Considering genetic relations, it is difficult to postulate a close connection between refractory inclusions and layer lattice silicate matrix. Oxygen isotope data support this. They show no genetic relationship between f.e. matrix and olivine separates of the Murchison CM-carbonaceous chondrite (Goswami and Macdougall 1983). Apparently both, the refractory inclusions and the matrix components were formed at different thermodynamic conditions and independent from one another. As has been shown by Christophe Michel-Levy et al. (1982) and Kurat et al. (1975), the more refractory inclusions were formed, at least in parts, at "violent" events. In contrast, the intermixture and compaction of both refractory and volatile species producing carbonaceous chondrites was a rather gentle process. Provided the density and collision velocity in the grain "mixture" were sufficient the existence of solid bodies which contain constituents with the described very different (in many cases even contrary) genetic properties points to a fast formation process of planetary parent bodies.

Physical implications:

Usually, investigations of meteorites under genetic aspects take place mostly from chemical and mineralogical-petrological points of view. Physical investigations are less frequent possibly because the genetic interpretation of the results is more difficult. Nevertheless, physical parameters can also provide additional constraints.

According to accepted theories about planetary formation processes the aggregation of millimeter-sized particles to planetesimals and greater planetary parent bodies is supposed. Concerning the growth mechanism of small grains, their mechanical properties are of special interest. This parameter depends on mineralogical and chemical composition, texture, bulk density, porosity, etc. Furthermore, the degree of consolidation of the constituents of the growing grains or planetesimals plays a significant role.

Yomogida and Matsui (1982, 1983, 1984) investigated these questions. They suggested that the initial inner constitution of chondritic bodies should be similar to that of solids consisting of dust grains which are loosely packed rather than to that of consolidated (rocky) materials. But, chondritic meteorites are consolidated, rocky-like, objects. Therefore, the grain aggregates must have experienced a compression to form a chondritic state. A suitable indicator to estimate the consolidation state is the thermal conductivity. Results of these experiments are shown in Figure 3.9.

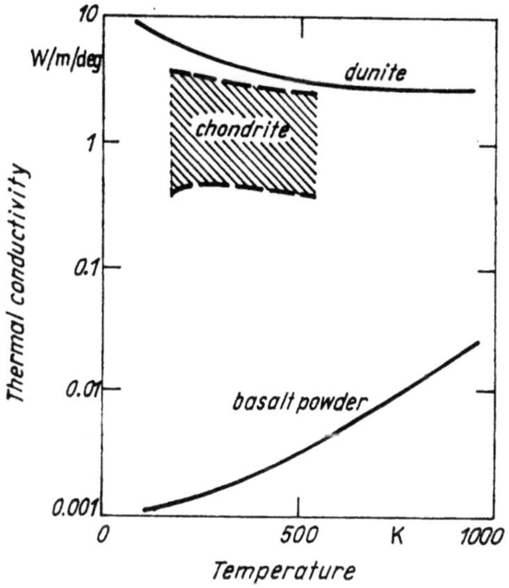

Fig. 3.9. Dependence of thermal conductivity on temperature for idealized dunite, chondrite, and basaltic powder (YOMOGIDA and MATSUI 1984).

The basaltic powder exhibited a porosity of 40% and a particle radius of 10 μm. Considering the thermal conductivity the measured values of compressed or consolidated dunite and chondrite are significantly greater than those of the basaltic powder. The authors attributed this to a sintering process. The porosity of chondrites is near or less than 20%. Therefore an appropriate mechanism is postulated reducing the porosity. This could be a sintering process at elevated temperatures. YOMOGIDA and MATSUI (1984) suggest that suitable temperatures can be reached by decay of long-lived radioactive elements. Other heat sources, as ^{26}Al, electrical induction by an early T-Tauri star or strong impact heating seem to be less appropriate. Further it is shown that differences in thermal conductivity between powder-like and consolidated material (Fig. 3.9) can change significantly the thermal evolution of small bodies. The maximum temperature reached in a body is nearly independent on the initial temperature and the diameter of the object. In contrast, it depends very sensitively on the temperature at which sintering begins. This behaviour results from better isolation properties of a loosely packed aggregat of fine dust particles. Therefore, long-lived radioactive elements are able to generate higher sintering temperatures in a greater time interval.

Besides thermal conductivity also elastic properties vary considerably with porosity. The elastic wave velocities correlate well with thermal conductivity. Furthermore, the porosity varies still with the petrologic type of the chondritic meteorites. Towards higher petrologic types, the porosity decreases. That means that a more intensive compaction is caused by rising temperatures. There is no analogous correlation for L-chondrites. Here secondary effects should have played a role. Impact processes should not have affected the porosity. After SIEGFRIED et al. (1977, 1981), the crack porosity increases only

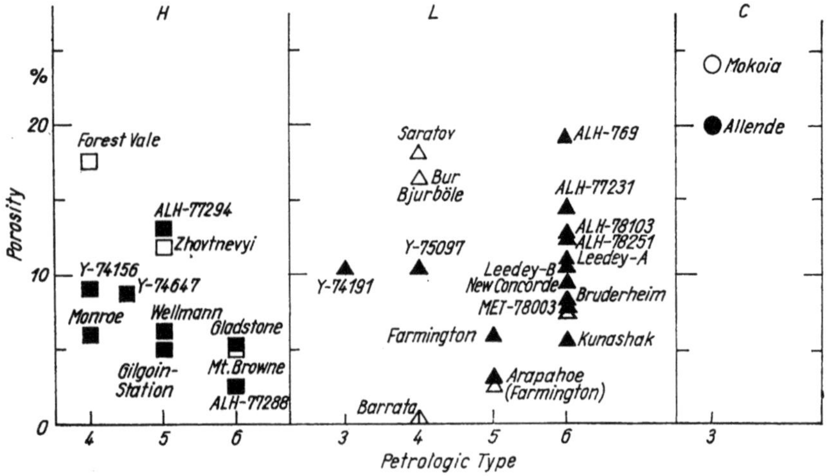

Fig. 3.10. Porosity versus petrologic type of H-, L-, and C-chondrites (Yomogida and Matsui 1982, 1983).

up to 1% at high velocity impacts with stresses between 0,2 and 2,0 GPa. The porosity versus petrologic type is shown in Figure 3.10.

Whereas the porosity sligthly decreases towards higher petrologic types at H-chondrites, there is no such trend at L-chondrites. Moreover, many of the L6-chondrites are characterized by higher porosities than the unequilibrated ones. The C-chondrites exhibit the highest porosities.

As described above, the knowledge of physical properties can contribute towards a better understanding of origin and evolution of preplanetary and planetary material. Particularly the results of the investigation of the porosity and the following effects on the thermal evolution of smaller planetary bodies are important. There is a clear hint that the first accretion of condensates yields highly porous "fluffy-type" planetesimals. Thus this process should have been a gentle one. The further evolution to "rocky" meteorite parent bodies was affected significantly by increasing temperatures. At this, probably longlived radioactive elements played a significant role. The distribution and density of pores and cracks shows also that the H- and L-chondrites underwent many different metamorphic events. The same characterization is applicable on the meteorite precursors surely.

Of course, only a few models are presented here briefly. But, it seems to be the real point of view that many mechanisms are responsible simultaneously for the formation of the known chondrite constituents and then for the preplanetary solids also. In this connection, a significant conclusion is that many scenarios allow the existence of presolar solids in the preplanetary material system which has undergone a manyfold of interactions. All that took place in very short time intervals.

3.5. The Composition of the Cometary Silicatic Component

Up to the exploration of Halley's comet was the generally consent that the most primitive material which we know are CI-carbonaceous chondrites and chondritic interplanetary dust particles (IDP's). So, in connection with the scientific expectations regarding the comet Halley experiments, it seemed natural to assume that a comet contains similar material or still more primitive matter.

In the result, the dust impact analysers (Kissel, 1986) carried by the spacecrafts VEGA 1, VEGA 2 and Giotto yielded information about the chemical properties of individual cometary dust particles. Essentially, there are three main classes of particles:

- particles, constituted of the elements C, H, O, and N,
- particles, constituted of the elements C, H, O, N, Mg, Si, Ca, and Fe. These particles might be compatible with the so called "Greenberg-particles", which consist of a volatile envelope (C, H, O, N-compounds) and a silicatic core (Fe, Mg, Ca, Si, O-compounds).
- particles, consisting one of silicatic components.

A comparison of the element ratios Mg, Fe, Ca, Al, Na, K, S and O to Si of these particles with the same element ratios of the CI carbonaceous chondrites and other terrestrial and extraterrestrial material is shown in Fig. 3.11. This comparison bases upon 32 spectra which represent silicatic particles only (experiment PUMA-1 of the spacecraft VEGA 1). The element ratios were compared with the ratios in CI-matrix, CI-average, interplanetary dust particles (IDP), Apollo 11/12 basalts, and terrestrial ultrabasites (Wäsch, 1986).

As a most recent result, Jessberger et al. (1988) published datas listed in Table 3.13. They point out the major rock forming elements within a factor of two have the same abundance pattern as the CI-chondrites and the sun. Very impressive is the fact that Halley's dust is enriched in the very light and volatile elements. This demonstrates that it is more solar-like, and hence probably more primitive solar system material than available as CI-carbonaceous chondrites. Mineralogically, Jessberger et al. (1988) speculate that the silicatic component might be dominated by more refractory silicates rather than hydrated layer lattice silicates. This speculation is based on the distribution of Fe/(Fe + Mg). In Halley's dust it ranged from 0 to 1 but with several regimes well separated. There are two groups with zero Fe and low Mg and two groups with Fe(Fe + Mg) around 0,2 and 0,5 respectively. In comparison, a layer lattice silicate IDP and the matrix of the Ci-chondrite Orgueil show Fe/(Fe + Mg)-ratios which are strongly peaked at 0,5. Contrary to this, an anhydrous IDP has a strong maximum at very low ratios, but covers thewhole range up to 1,0. Thus Halley's dust resembles an anhydrous IDP rather than matrix material of the Orgueil chondrite. Therefore, Jessberger et al., (1988) speculate that among layer lattice silicates also anhydrous silicates may occur.

After all, this speculation appears to be supported by IR-spectroscopic measurements (McKeegan et al., 1987). According to them, the composition of Halley's dust is 56% olivine, 36% pyroxene, and 8% layer lattice silicates.
If these results are true anhydrous silicates like f.e. olivine or pyroxenes appear to be "more primordial" than hydrated layer lattice silicates.

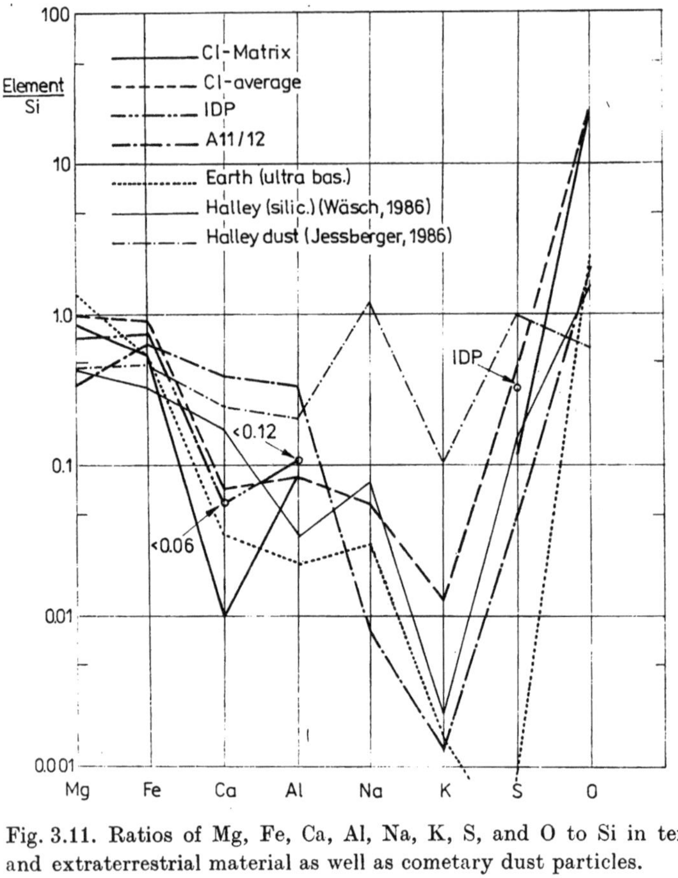

Fig. 3.11. Ratios of Mg, Fe, Ca, Al, Na, K, S, and O to Si in terrestrial and extraterrestrial material as well as cometary dust particles.

Summarizing all today known (still rough) mass spectrometer results of the comet Halley missions, the nature of the cometary silicatic dust component seems to be very close to relatively undeveloped preplanetary material of the solar system.

That apparently applies to the mechanical construction of the dust particles too. The densities lie in the range of 0,01—0,4 g/cm³ (mean 0,1 g/cm³) which is also close to values derived from meteor phenomena. This confirms the fluffy nature of cometary particles and on this base a gently agglomeration process of such particles.

Table 3.13. Average atomic abundances of the elements in Halley's dust and Type 1 cabonaceous (CI)-chondrites (from JESSBERGER et al., 1988)

Element	Halley's dust	CI	Halley/CI
H	2025 ± 385	492	4.1
C	815 ± 165	70.5	11.6
N	42 ± 14	5.6	7.5
O	890 ± 110	712	1.3
Na	10 ± 6	5.3	1.9
Mg	= 100	= 100	= 1.0
Al	6.8 ± 1.7	7.9	0.9
Si	185 ± 19	93.0	2.0
S	72 ± 23	47.9	1.5
K	0.2 ± 0.1	0.35	0.5
Ca	6.3 ± 1.9	5.68	1.2
Ti	0.4 ± 0.2	0.22	1.9
Cr	0.9 ± 0.2	1.25	0.7
Mn	0.5 ± 0.2	0.88	0.6
Fe	52 ± 9	83.7	0.6
Co	0.3 ± 0.2	0.21	1.2
Ni	4.1 ± 2.1	4.59	0.9

3.6. Some Open Problems

Some of the most significant uncertainties are:

— Condensation and nucleation of preplanetary solids from a gas phase.
We do not know the thermodynamic conditions and reaction kinetics in a cooling solar nebula or, f.e. in expanding supernova shells.

— Decay reactions of radioactive elements.
Where and how the isotopic ratios were established (during stellar processes or by reactions within the Solar System or both)?

— Dating of material classes and individual minerals, components, etc.
The experimental methods used up today allow to separate material ages up to few tens of million years. But, basing on our actual knowledge, it seems to be necessary to have a resolution of different ages, at least of the order of one million years (or better).

- The temperature regime of the Solar System.
  In this connection parameters as condensation sequence, modification of eventually injected solids of stellar processes, element distribution, and material separation along the Solar System temperature gradient, reheating, revaporizing, etc. are significant.
- The influence of supernovae, novae, etc. on origin and evolution of the Solar System.
  It is interesting to know whether stellar processes are as well a trigger mechanism for the origin of the Solar System as a considerable source of the material pattern of the Solar System. And last in general,
- What are the conditions and the time of origin of the preplanetary material (genealogy of solid matter)?

The problem of the origin of solid preplanetary matter can be investigated also from other points of view. Such a case is presented in section 3.7. At this on approach is tried which bases upon the element abundances in the crust of the Earth and their relation to cosmic abundances. By the developed model it is taken into account that the ubiquitous hydrogen dominates all first step chemical reactions. The resulting effect on the origin and composition of the Earth is considered.

3.7. Geochemical Evidence of Early Protoplants
(by R. Gottfried)

A different approach to the origin of a part of solid matter in the Solar system comes from the comparison of the abundances of elements in the crust of the Earth and their average "cosmic abundance", as it is inferred from primitive C1-chondrites (Hahne, 1983) or from astrophysical investigations (Cameron, 1968). Even a first impression it becomes clear that some of the elements are highly enriched (as Thorium, Barium, Lanthan and Tantal). The concentration of these elements in regions of the Earth's crust has been given by Turekian and Wedepohl (1961) and Vinogradov (1962). The resulting masses in the crust are described by Mason (1966), Wedepohl (1969) and Ronov and Jaroshevskij (1976). They can be used to estimate their total mass in the early Earth body in the following way.

The cosmic abundance of elements is given with a relative number of atoms with the normalization to $S_i = 10^6$. The relative mass m_{rel} of an element is then given by the product of abundance and mass of the atom of the element. The total mass of all elements with cosmic abundance is then the sum of all products. This is given with $3.53 \cdot 10^{10}$ if the above mentioned normalization $S_i = 10^6$ is used. Consequently, the quotient q_{el} of the contribution of an element is given then with

$$q_{el} = \frac{m_{rel}}{3.53 \cdot 10^{10}} . \qquad (3.7.1.)$$

If now the mass m_{el} of this element in the Earth is known, the total mass of the "proto-Earth" (m_{proto}) can be determined from

$$m_{proto} = m_{el}/q_{el}. \quad (3.7.2.)$$

Table 3.14. Original mass of the Earth

enriched Element	cosm. abundance $Si = 10^6$	content in the crust (ppm)	$q_{Element}$	mass of Element (kg)	M_{proto} = calculated total mass
Ba	4.7	700	$1.83 \cdot 10^{-8}$	$2 \cdot 10^{19}$	$1.1 \cdot 10^{27}$
Th	0.034	15	$2.23 \cdot 10^{-10}$	$4 \cdot 10^{17}$	$1.8 \cdot 10^{27}$
La	0.36	60	$1.4 \cdot 10^{-9}$	$2.6 \cdot 10^{18}$	$1.8 \cdot 10^{27}$
Ta	0.022	4	$1.13 \cdot 10^{-10}$	$1.3 \cdot 10^{17}$	$1.2 \cdot 10^{27}$
all elements	$\Sigma H \cdot A = 353 \cdot 10^8$	10^6	1	average = $(1.5 \pm 0.4) \cdot 10^{27}$	

The resulting values are given with Table 3.14. The average value of

$$m_{proto} = (1.5 \pm 0.4) \cdot 10^{27} \text{ kg} \quad (3.7.3.)$$

indicates an early massive proto-Earth. This value has to be understood as a minimum value in that sense that not all of the atoms of the used elements can have had accumulated in the crust. On the other side this has to be understood as a maximum, if the preplanetary disk in the Earth-forming region has lost volatiles in a yet earlier phase (producing on this way a deviation from the original cosmic abundance). Nevertheless, as an order of magnitude estimation the mass of proto-Earth can be assumed to be given appropriately by $m_{proto} \approx 10^{27}$ kg.

This model of a proto-Earth, having a more or less "cosmic abundance" of the elements, is necessarily connected with a "hydrogen-dominated" chemistry. Resulting conclusions shall be discussed shortly in the following, but this is a yet open field for future "planeto-chemistry". The discussed conclusions have to be seen only as first estimations. Partially they could be confirmed by comparison with properties of Jupiter and Saturn, which in some sense could be comparable to protoplanets as proto-Earth.

In principle, in chemical reactions the dominating hydrogen has a reducing activity, leading to the elementary state (of the involved elements) or it produces hydrids. An early internal structure of the protoplanet could be characterized therefore by a core of metals or metallized substances followed upwards by "mantle" of saline hydrides and fluid or gaseous hydrids (where hydrogen is bound covalently). The "atmosphere" of this protoplanet should consist mainly of hydrogen and helium. This structure model could be related with observed facts in the present crust of the Earth.

Obviously, there is a depletion of elements in the crust, which are able to form hydrides with low density or gaseous hydrides or which can be reduced to the elemental state (sinking down to the core). On the other side, elements are enriched relatively, which can form saline hydrides.

A parameter, appropriate to describe this, is the socalled "electronegativity" of elements. Fig. 3.12 gives the ratio (QE) of the electronegativity of hydrogen and of the other elements in correlation with the above mentioned deviation Af of crustal abundances from cosmic abundances of elements. Obviously, hydrogen acts as anion for large QE, the resulting hydrides are saline hydrides. The corresponding elements are enriched relatively. For low QE hydrides do not form or have the character of metal (if the elements are metallic), while the other elements form (by covalent bound with hydrogen) gaseous hydrides which behave as volatiles.

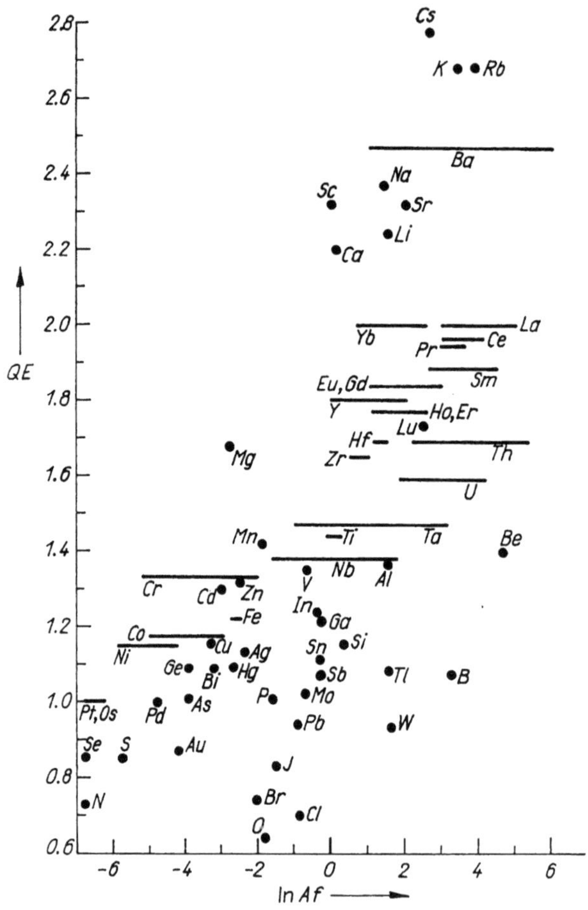

Fig. 3.12. Relation between the deviation Af and the electronegativity QE of hydrogen and other elements.

A further parameter, appropriated to describe the chemical activity of elements with respect to hydrogen is their "ionization-potential "I" (GOTTFRIED, 1987). Using 41 elements GOTTFRIED (1987) derived quantitatively (with a correlation coefficient of 0.85).

$$I^{+} = -0.4 \ln Af + 6.6 . \qquad (3.7.4.)$$

A similar relation follows if instead of the deviation Af from cosmic abundances that of the concentration in C1-meteorites is used, or if modifying effects of magma-differentiation in the crust are taken into account. This relation is a strong argument in favour of the above formulated hypothesis of the essential influence of hydrogen on the evolution of the proto-Earth.

Summarizing this geochemically motivated first discussion of the chemistry in the protoplanetary cloud of reducing hydrogen it can be stated (GOTTFRIED, 1979, 1987), that essential properties of solid planetary matter (Luna, meteorites, Earth crust) can be understood as to be caused by the reducing influence of protoplanetary hydrogen. Of course, this does not contradict the importance of other (láter) differentiating and transport processes, but an essential first trend in forming solid planetary matter seems to be well described by taking into account these effects of the reducing protoplanetary hydrogen atmosphere. This model leads to the following conclusions about the structure of the proto-Earth (as an example for a proto-planet):

- metallic or metallized core-mantle region including hydrides (sinking down due to their greater density),
- thin crust of saline hydrides (of lower density),
- a fluid layer of hydrides of non-metallic elements and liquidized gases,
- a (gaseous) hydrid-hydrogen-helium atmosphere.

A further geochemically motivated discussion of this new approach to protoplanets seems to be a necessary future task.

4. Evolution of Planetary Bodies and their Interiors

H. Stiller; S. Franck; U. Schmit; I. Orgzall

In this part an attempt is made to describe the evolution of planetary bodies after they had formed. This formation process will be outlined in more detail in part 5.

The starting-point of the considerations presented here is the description of the present interior structure. This clearly indicates that large-scale differentiation processes had to operate to produce the highly differentiated structure from a — as it is assumed here — primary homogeneous body. In the case of the terrestrial planets this process is the formation of the core. Based on simple models the energetics of the core formation and related phenomena, as the generation of magnetic fields, are discussed. Another section deals with the thermal evolution of the planetary interior, discussing temperature profiles during planetary growth and their compatibility with different planetogonie approaches.

The group of giant planets is treated in an analogous manner. The present structure as product of long-during evolution processes is considered and with a detailed analysis of the phase diagram of the most abundant element (hydrogen) and related thermodynamic considerations the models of the interiors are constructed.

4.1. General Aspects of Planetary Evolution

The formation of the present interior structure of the planets is closely related to the problems of planetary cosmogony. One approach to this formation process is Schmidt's hypothesis about the accretion of small preplanetary solid particles (planetesimals) to form the planets. An alternative planetogonic model is that of protoplanets as predecessors of the present planets.

Within the frame of accreforme models the structure of the resulting planets is characterized by two different approaches. If the planetesimals contain both iron or iron compounds and silicates in nearly chondritic abundances the evolution leads to a primary homogeneous planet with only smallscale inhomogeneities in the range of the size of planetesimals. This is called homogeneous accretion. The present differentiated structure is obtained by chemical and physical processes in further planetary evolution.

The other model, the inhomogeneous accretion, is based on the assumption of large-scale physical or chemical processes of separation of iron and silicates in the preplanetary nebula. These separation mechanisms are founded on differences in plasticity and elasticity, in magnetic properties or in the condensation temperatures from the gas phase (ANDERS, 1972). OROWAN (1969) suggested such processes. So metallic particles would stick together when they collide since they can absorb kinetic energy by plastic deformation. Silicates are brittle and would break on collisions. This behaviour is observed in several experiments. As LELIWA-KOPYSTYŃSKI (1984) and LELIWA-KOPYSTYŃSKI et al. (1984) point out, from sticking experiments it follows, that target mass increase may by expected for ductile materials, especially for metal—metal collisions or rock—rock impacts at elevated temperatures near the melting point so that the material is in the region of a brittle—ductile transition. But this process depends on a lot of parameters as particle size, relative velocity, impact angle etc.

So the primarily formed planetesimals will consist of a large iron fraction. These planetesimals accumulate first and build the planetary core. Later the silicates condense from the nebula followed by several volatile elements. The

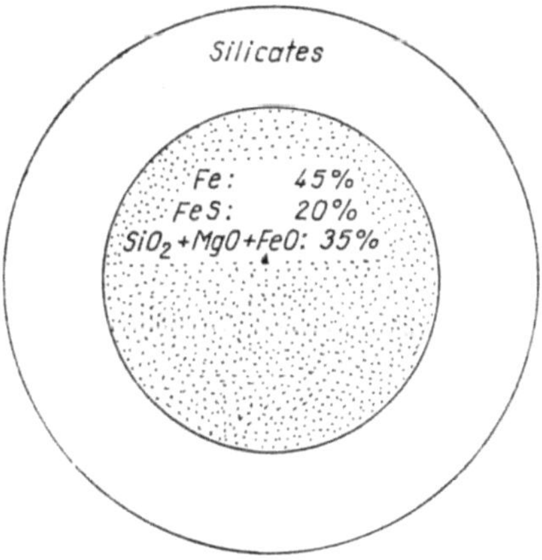

Fig. 4.1. Model of a primary inhomogeneous Earth (from MAKALKIN, 1980).

accumulation of the planetary embryos continues with the silicate fraction forming the mantle. This model was discussed for instance by MAKALKIN (1980) in application to the Earth. Due to their physical properties iron-rich dust particles grow bigger than the silicate-rich and for this reason settle faster down to the midplane of the disk. So these planetesimals accumulate to form a primary core with a radius of about 4600 km and nearly 50% of the whole Earth's mass. The silicates arrange around this iron-rich fraction (cf. Fig. 4.1). After the formation differentiation between iron and silicates begins, but different from other models this process affects only the core region. Such models are discussed for instance in connection with considerations about the volatile content, especially H_2O, of the Earth (LANGE and AHRENS, 1984).

A fully differentiated model with Fe—Ni core and silicate mantle is proposed for example by TUREKIAN and CLARK (1969). The constituents condense in the sequence iron, nickel followed by magnesium and iron silicates, alkali silicates, various metals up to iron sulfide, Fe_3O_4, volatile trace elements, organic compounds, hydrated silicates and inert gases from the solar nebula and accrete in the same sequence. So primary a Fe—Ni core occurs and the other materials assemble around forming the layered structure observed at present. Fig. 4.2 illustrates this process in a simplified manner. Arguments supporting this hypothesis are discussed and refuted by VITJAZEV (1982).

The main difficulty that is involved in the models of inhomogeneous accretion is the explanation of the occurrence of the latter differentiation, especially the origin of a liquid outer core.

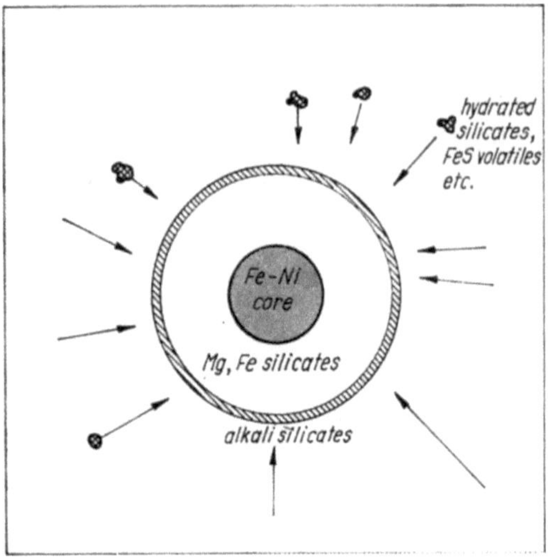

Fig. 4.2. Inhomogeneous accretion of the Earth from materials in the sequence of their condensation from the solar nebula.

In the following models of homogeneous accretion are discussed more in detail and the differentiation process of these planets is taken into account in sections 4.2.2. and 4.3.2. respectively.

The composition of the planets devides those roughly into two groups: the terrestrial planets composed from "rocky" matter and the giant planets from gaseous and "icy" matter. The "rocky" matter consists mainly of iron, iron compounds such as FeS and FeO, oxides, silicates of several elements as Na, K, Ca, Fe, Mg and Al with several crystal structures, for instance olivine, spinel and other structures, depending on pressure and temperature and therefore depth, of trace and radioactive elements and of volatile components like H_2O, CO_2 and others (cf. for instance LEWIS, 1972; ZHARKOW and TRUBITSYN, 1980; WÄSCH, 1983). Roughly these planets were formed by materials in nearly chondritic abundances (cf. part 3.).

The gaseous and "icy" matter consists of volatile elements and compounds as H_2, CO_2, H_2O and CH_4, NH_3 in condensed, that means liquid or solid, and in gaseous phases. The most abundant elements in these planets are H_2 and He. The hydrogen is partially in a metallized state (see section 4.3.1.). This matter is accompanied by a small rocky fraction so that it may be compared to a dirty snowball as it is known from cometary investigations (FINK and IP, 1983; see also ZHARKOV and TRUBITSYN, 1980).

The differentiation process that follows the accretion or, as some models assume, that accompanies this process after the radius of the planet has reached a certain critical value is determined by thermal parameters. It is closely connected with the primary temperature distribution because the position of this curve relative to the melting curves of the several components is essential for the redistribution of the masses in the planetary interior. Another important quantity is the viscosity of molten and unmolten material. This quantity is strongly temperature-dependent.

The basic principle that rules the formation of the planets as well as the differentiation process is the principle of conservation of energy. This gives the possibility of evaluating the primary temperature distribution (cf. ZHARKOV and TRUBITSYN, 1980; RANSFORD, 1982).

$$\frac{\varrho\gamma M(r)}{r}\,\mathrm{d}r = \varepsilon\sigma[T^4(r) - T_B{}^4]\,\mathrm{d}t + \varrho c_P[T(r) - T_B]\,\mathrm{d}r \tag{4.1}$$

where

ϱ — density of the accumulating particles;
γ — gravitational constant;
$M(r)$ — mass of the growing planet with radius r;
σ — Stefan-Boltzmann's costant;
ε — emissivity;
c_P — specific heat at constant pressure;
T_B — basic temperature of the accumulating particles.

The left hand side of equation (4.1) describes the gravitational energy released from the impact of the planetesimals. This energy is partly radiated back into space (first r.h.s. term of (4.1)). The other part is transformed into thermal energy, so that the material warms up (second term). If the material is partially molten on the r.h.s. of (4.1) one has to include another term involving the latent heat. But due to its analytical structure it is possible to combine this term with that of the specific heat. The effect of short-living radionuclides as for instance ^{26}Al may be taken into account modifying the basic temperature T_B. Other theories involving more and complicated processes that describe the planetary evolution and therefore the thermal behaviour more realistically, have been worked out by SAFRONOV (1969) and KAULA (1979 and 1980).

For the further evolution of the planetary interior several processes that influence the thermal state significantly have to be considered. First of all the radioactive decay of long-living isotopes as for instance ^{238}U, ^{235}U, ^{232}Th and ^{40}K. Other processes that affect the energy balance and the thermal profile are heating due to gravitational energy, including compressional heating and the energy released at the core formation, and tidal heating (HOREDT, 1980a). Convection in several regions of the planetary interior and heat conduction act on the energy balance as well.

In the following chapters the two groups of planets will be discussed in more detail: at first the terrestrial planets and their mechanisms of differentiation and core formation including a detailed analysis of thermal processes. These models are appropriated to discuss conclusions from the "accrebonal approach" to planetogony. Secondly, there is the group of the giant planets. Especially the hydrogen phase diagram of the metallic phase and the resulting consequences for their structure are discussed. The origin of these planets can be used as a starting point for that of protoplanets. Therefore, they could give information on the physics of such a protoplanet as a preplanetary state.

4.2. The Terrestrial Planets

There are known a lot of facts about the terrestrial planets Mercury, Venus Earth and Mars. Well structured models for these planets were developed The Earth offers the possibility of direct observations, for instance by seismic sounding, drilling etc. The other planets have been explored by many automatic spacecraft missions and observations from Earth. But of course not all features are well understood, so a lot of work has to be done in the future. Even the knowledge about the Earth's interior is far away from an adequate description of the complex behaviour of matter under high pressures and temperatures and of the resulting interior structure.

The main components of the terrestrial planets are materials of rocky type as silicates, other minerals, iron and iron alloys, oxides, other metals and

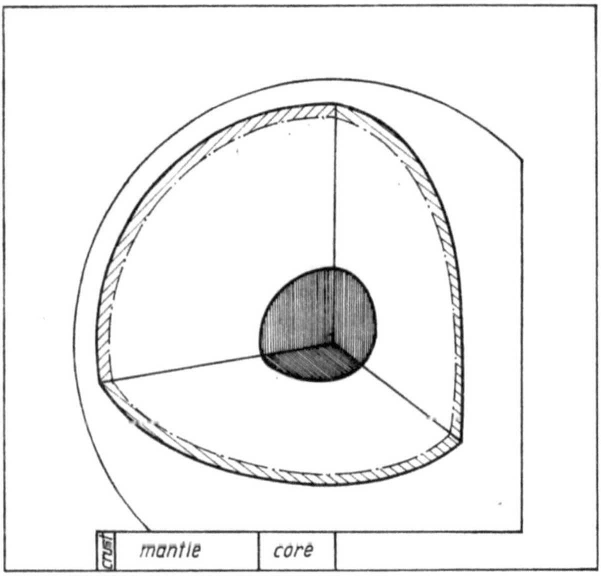

Fig. 4.3. Main structure elements of a terrestrial planet.

elements, and a small fraction of volatiles. This fraction depends on the accretion scheme (for instance for the water content see LANGE and AHRENS, 1984) and on the distance between Sun and growing planet. Exact element abundances are found for instance from ZHARKOW and TRUBITSYN (1980) or in part 3 of this book.

The structure of these four planets can be divided roughly into two main constituents: core and mantle (see Fig. 4.3).

The core consists of iron with alloying elements. Another hypothesis should be mentioned here. If the pressure is sufficiently high, it is expected that the matter is metallized. A crude estimation of that pressure by STEVENSON (1980) results in $p \simeq 10^{11} \ldots 10^{12}$ Pa. This is a pressure that occurs in all planets (at various distances from the centre) and so a metallization should be possible.

A hypothesis about the core's composition from metallized silicates was developed by LODOTCHNIKOV (1939) and RAMSAY (1949). But this metallization of silicates would lead to melting and a density discontinuity, and there is no experimental evidence (cf. STILLER et al., 1980) neither for this nor for the modified theory by ARTJUSHKOV and LJUSTICH (see LJUSTICH, 1974), that in a two-component system only the denser component is metallized, what would reduce the discontinuity.

The core is structured into a solid Fe—Ni inner core and an outer liquid core of Fe with light alloying elements as sulphur for Venus and Earth. For Mars the same is assumed, but depending on the model, the core possibly could be solid too as it is assumed for Mercury.

The mantle consists of silicates and oxides as FeO in various abundances. In that region convection takes place. The uppermost layer, the lithosphere, determines the tectonic activity. So called one-plate planets as Mars and Mercury are inactive. The major planets Earth and Venus have thin lithospheres and especially the Earth shows an active plate tectonic behaviour. Tectonic

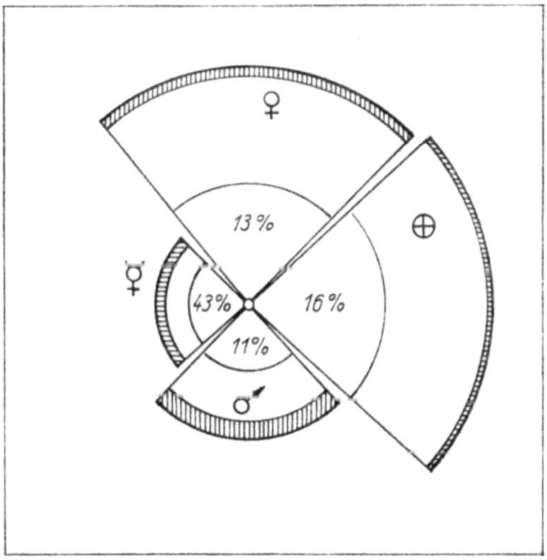

Fig. 4.4. Comparison of the four terrestrial planets: the number gives the ratio core/whole planetary volume; dashed: lithosphere (☿ — Mercury; ♀ — Venus; ⊕ — Earth; ♂ — Mars).

activities are assumed for Venus too. Fig. 4.4 illustrates the size of the planets, the ratio between core and mantle and the thickness of the lithosphere.

LEWIS (1972) shortly characterized the terrestrial planets:

Mercury: a massive iron core, surmounted by a silicon-deficient and FeO-free mantle and a potassium-free crust.

Venus: a massive iron core, a mantle containing about 1% FeO, and a potassium-bearing crust.

Earth: a core mainly composed of an iron-sulphur melt, a mantle containing about 10% FeO, and a potassium-deficient crust (relative to uranium, etc.) containing a trace of water.

Mars: a small core of iron sulphide, a very FeO-rich mantle ($\sim 50\%$), and a potassium-rich crust, very rich in chemical bound water.

Because of the similarity of these planets to the Earth the well developed and accepted models may be applied to Mercury, Venus and Mars with small corrections for the specific differences in the appropriate boundary conditions. In the next sections based on a consideration of the present state the evolution of the planetary interior, especially the core formation, is discussed together with a detailed energy analysis with special regard to the dynamo action.

4.2.1. Present Structure

The structure of the planetary interior, as it is observed directly for instance by seismic studies or as it is concluded from secondary data as period of rotation, dynamic figure or the magnetic field and corresponding models, is the result of a long-during process beginning with the growth of the planet from the pre-planetary disk. This process is constrained by many different aspects as the composition of the protoplanetary matter at the planet's distance from the Sun, the growth mechanism and essentially the growth time that determines the primordial temperature by the energy balance equation (1), the abundance of radioactive elements, the scenario for the core formation process, energy release and energy flow (convection, conduction, radiation) and therefore the cooling of the planetary interior and many others not mentioned here.

The main structures are, as already mentioned, crust, mantle and core. Fig. 4.3 gives a general model for all terrestrial planets. The specifications for the Earth are described in Fig. 4.5. The layers indicated there are structured themthelves into smaller layers. That means, that they are not homogeneous, as may be deduced from this figure, but they are horizontally and laterally inhomogeneous. The boundaries between the layers are characterized by phase transitions. So the crust-mantle boundary, the so-called Mohorovičič discontinuity is a chemical discontinuity that formed during the evolution of the mantle due to differentiation processes. The core-mantle interface is also such chemical phase boundary between the silicate-rich mantle and the iron-rich core.

The inner-core boundary is characterized by the transition solid-liquid, but it can also be interpreted as a chemical boundary between core material that is depleted in lighter components as sulphur (solid inner core) and enriched

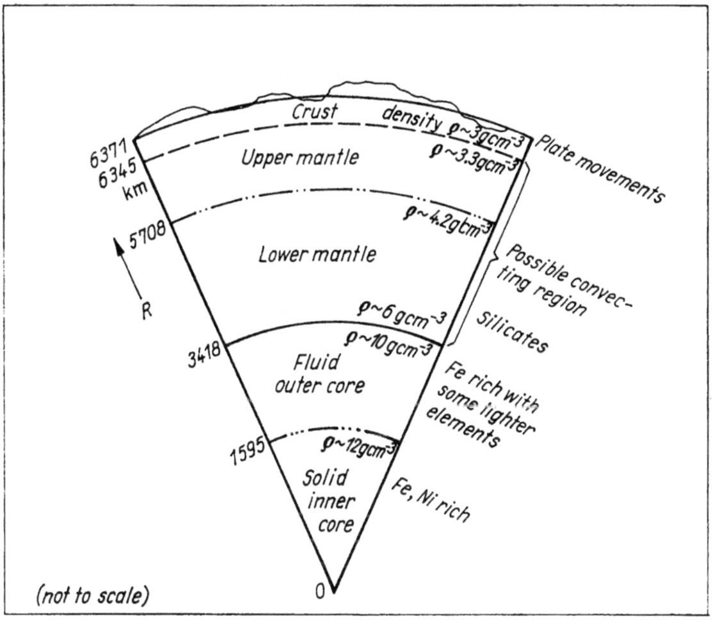

Fig. 4.5. The Earth's internal structure as deduced from seismic studies (from COLE, 1981).

material (liquid outer core). This differentiation is due to solidification processes in a many-component system (for a detailed discussion see section 4.2.2.). These processes must have an influence on the convection and the released energy can contribute to power the geodynamo (BRAGINSKI, 1963).

From seismic velocity distributions three important discontinuities may be concluded for the mantle: a low velocity zone in a depth of about 100 ... 200 km (LVZ) and two zones with an anomalous velocity increase in a depth of 400 to 430 km and 650 ... 740 km, respectively. The LVZ is interpreted as region with partially molten material. It is of great importance for plate tectonic movements of the overlying lithosphere. Basing on high-temperature and high-pressure research RINGWOOD (1981) interprets the 400-km-discontinuity as phase boundary between α- and β-phase of olivine $(Mg, Fe)_2SiO_4$. The 700-km-discontinuity should be a phase boundary too. Below this interface one should expect minerals with post-spinel structure (perovskite, oxides with NaCl-structure) with sixfold coordination of oxygen around silicon (see also LOPER, 1985).

For a theory that connects seismic discontinuities with thermodynamic considerations, that especially takes into account the behaviour of the thermodynamic functions and of an — in the framework of Landau-theory for phase transitions (LANDAU and LIFSHITS, 1971) introduced — order parameter in the region near to those transitions and that results in three different types of the behaviour of seismic velocities at phase transitions see for instance STILLER and FRANCK (1980). With these types it is possible to explain the discontinuities in seismic velocity distributions and experimental results from ultrasonic investigations as well.

The mantle consists of a mineral assemblage and so the phase boundaries will not be sharp. They are widened over a certain region.

For quantitative estimations of relevant physical parameters DZIEWONSKI et al. (1975) worked out a standard model of the Earth, the so-called Parametric Earth Model (PEM). This model with three modifications — the continental, oceanic and averaged form — reflects in a mathematically simple way the structure and properties of the Earth's interior. Figure 4.6 illustrates the

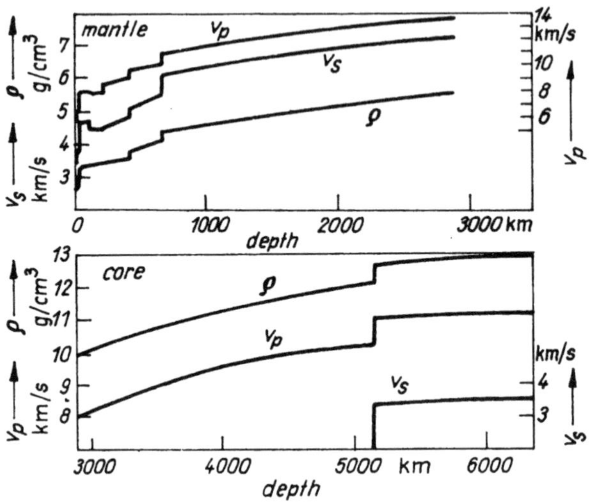

Fig. 4.6. Continental PEM (DZIEWONSKI et al., 1975).

continental model with appropriate velocity and density structures (for the other models the reader is referred to the mentioned literature). The discontinuities are clearly indicated. STACEY (1977) connected this model with thermodynamic considerations and so a complete model of the Earth's interior was developed. This radial symmetric model was improved by several authors and DZIEWONSKI and ANDERSON (1984) introduced a three-dimensional Earth model.

For the other planets those models were calculated too, for instance for Venus by ZHARKOV (1983). Of course, the best constrained models are those for the Earth because of a lot of available data and observations. The models for the other planets are rather rough, basing on many suppositions and they can reflect only some aspects of the complex behaviour and one always should keep in mind that these models themselves are based on model assumptions.

For Mercury, Venus and Mars the present state is shown in Fig. 4.7a–c as a result of evolutionary processes.

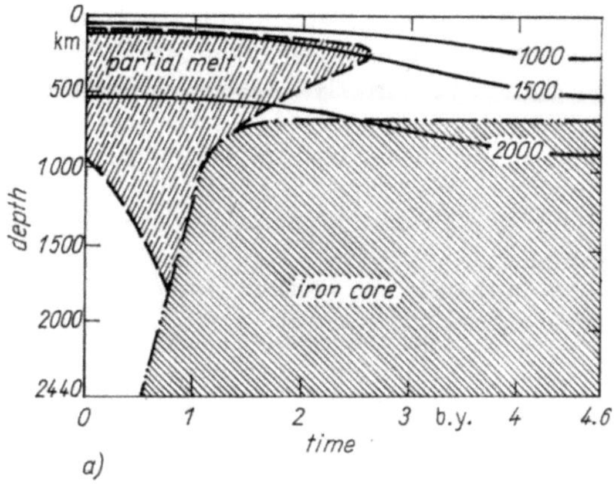

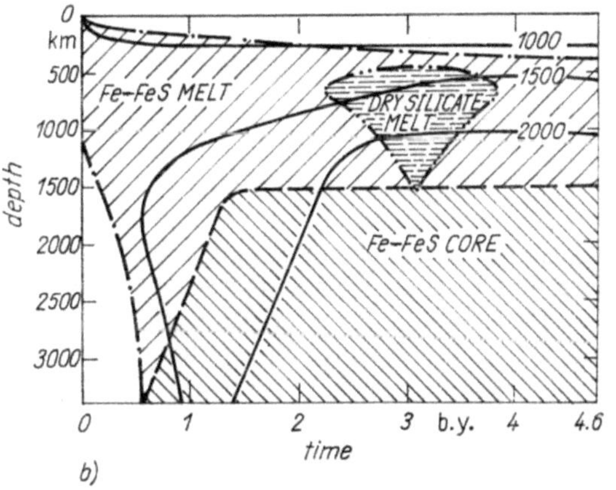

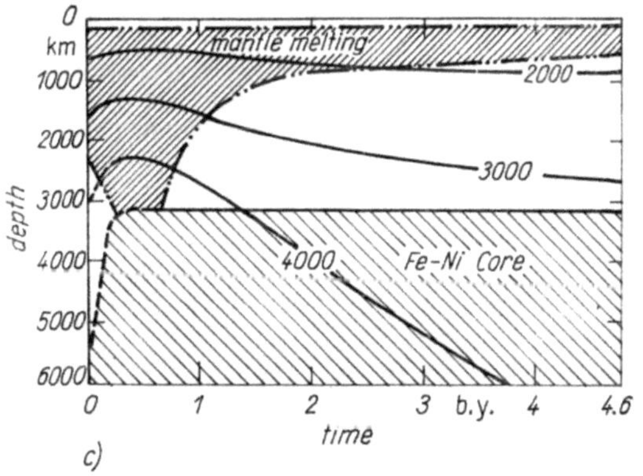

Fig. 4.7. Evolution of three terrestrial planets; isotherms in °C (from Toksöz et al., 1978).

a) Mercury b) Mars c) Venus

The uniform structure consisting of iron core and silicate mantle is evident but also the different evolutionary schemes. The terrestrial planets possess a magnetic field. It is expected that the not yet directly measured Martian field is weaker than that of Mercury.

Several processes are discussed to power such planetary dynamos. As it is known from the Earth, the magnetic field results from magnetohydrodynamic processes in the outer core. To establish a stationary dynamo a power supply is required to overcome the dissipation.

One conceivable source of energy is thermal convection. But thermal convection is only possible at hyperadiabatic temperature gradients and the efficiency is low.

The precession-driven dynamo (MALKUS, 1963) works due to currents in the core caused by precession of the rotational axis. The rotational energy powers the dynamo. But some authors opposed and calculated that the released energy is to small for supporting the dynamo.

BRAGINSKI (1963) proposed chemical differentiation at the inner-core boundary as the cause for power supply. Such effects are often observed at solidification of multi-component systems. At the phase boundary light elements are released, what leads to a convective instability and is followed by chemical convection. This is discussed in more detail in section 4.2.2. But always dynamo action requires an electrically conductive moving fluid in a rotating system.

If the iron core would be partially molten, so that motions, as for instance convection, could take place this would be a possible source for the origin of the magnetic field. But it is necessary for instance for Mercury to include internal heat sources within the core to secure partial melting. This problem is not solved yet.

A partially molten core is assumed for Earth and Venus with a liquid outer core (for models of planetary magnetic fields and appropriate data cf. RUSSEL, 1980).

For Mars it is yet difficult to decide between a small molten shell around a solid inner core or a fully solidified core (YOUNG and SCHUBERT, 1974).

The differences in the evolution of planets are due to the size (especially the radius), the composition, energy release by core formation and other relevant planetogonic parameters. These differences also result in different tectonic and volcanic activities of these planets.

The Earth is known as an active planet with large-scale plate tectonic movements and active volcanoes. These processes are connected with vigorous convection in the mantle. Mars is expected to be an inactive planet with active periods in the past, where the surface structures developed. For Venus it is believed that many surface structures are relatively young and probably of tectonic origin. The surface seems to be tectonically active yet (for the hypothesis of plate tectonics see BRASS and HARRISON, 1982).

The thermal behaviour of the planet is calculated from a system of equations as for instance the equations of hydrostatic equilibrium, energy balance, mass and moment of inertia for a spherical shell and for heat flow (BAUMGARDNER and ANDERSON, 1980).

The most important equation which has to be taken into account is an appropriate equation of state for the various materials (cf. for instance LELIWA-KOPYSTIŃSKI and TEISSEYRE, 1984). The results are temperature-radius and pressure-radius profiles for the different planets (cf. ZHARKOV and TRUBITSYN, 1980). Figure 4.8 shows such a p-T-r diagram for the present state of the Earth's interior (combined from DZIEWONSKI et al., 1975 and STACEY, 1977).

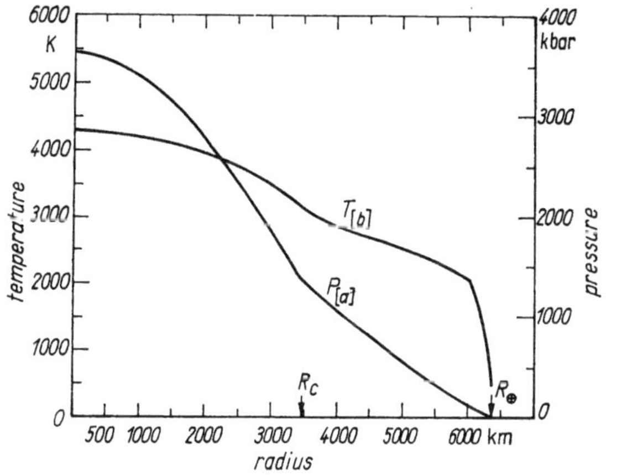

Fig. 4.8. Pressure and temperature inside the Earth — continental model (DZIEWONSKI et al., 1975 (a); STACEY, 1977 (b)).

4.2.2. Differentiation of the Planetary Interior

To construct formation models of the present planetary structure from a primary undifferentiated body it is necessary to consider firstly the temperature profile of the growing planet and secondly the melting curves of the various important constituents. From this, one can conclude where differentiation starts and how it proceeds.

The initial temperature profile, the profile at the end of the accretion can be derived from (4.1). Several calculations were carried out by SAFRONOV (1969 and 1972); TOKSÖZ et al. (1978); VITJAZEV and MAJEVA (1980); CORADINI et al. (1983) and other authors. These profiles depend on details of the accumulation process. Several different heating processes have to be taken into account. Release of gravitational energy connected with the growth of the planetary body is the most intensive heat source. The energy is liberated at the surface if small particles are considered to accrete and at a certain depth of several tens of kilometres for larger planetesimals. So a difference in heating efficiency for small ($r < 1$ km) and large ($r > 1$ km) bodies occurs (CORADINI et al., 1983) within the frame of accretional planetogonic models. A substantial part of energy is radiated back into space. The rest is stored in the interior. The surface temperature can be estimated by the Stefan-Boltzmann-law. For the storage the penetration depth of the "projectile" and the depth distribution of heat deposition are important. Additionally to equation (4.1) for quantitative calculations the heat transport equation has to be considered (CORADINI et al., 1983):

$$\frac{\partial T}{\partial r} = \left[\frac{1}{r^2}\right]\left[\frac{\partial}{\partial r}\right]\left[r^2 k\left(\frac{\partial T}{\partial r}\right)\right] + \left[\frac{E}{\varrho c_P}\right] \tag{4.2}$$

where T is the temperature, c_P specific heat at constant pressure, k thermal diffusivity and E the energy generation rate. A more general form is given by SAFRONOV (1972). An essential part of E is the energy E_I due to impact.

Another part E_R is due to the heat generation by radioactive decay. Important isotopes are ^{238}U; ^{235}U; ^{232}Th and ^{40}K. The energy release is given by

$$E_R = \sum_i A_i C_i \exp(-\lambda_i t) \tag{4.3a}$$

where the initial concentration C_i from about 4600 Ma ago is given by

$$C_i = C_{i0} \exp(\lambda_i \tau^*) \tag{4.3b}$$

with $\tau^* = 4600$ Ma, C_{i0} — present average concentration of isotope i, λ — decay constant and A — energy release per unit mass. The summation has to be taken over all radioactive isotopes. SAFRONOV (1972) estimates the heating due to this process to be small. It yields only a temperature increase of about 200 K in the Earth's centre.

The third effect to be included into the considerations is the temperature rise due to adiabatic compression. The new layers of impacting planetesimals cause an increase of the radius and therefore pressure increases in the interior. The adiabatic temperature profile is for $\varrho =$ const.

$$T(r) = T_0 \exp\left[\left(\frac{2\pi}{3c_P}\right)\alpha\varrho\gamma(R^2 - r^2)\right] \tag{4.4}$$

where T_0 is the temperature at $r = R(t)$ and α the volume coefficient of thermal expansion. The temperature is always higher than the minimum profile given by equation (4.4). The compressional energy assuming an equation of state of the form $P = a\varrho^n + b$ for the Earth's interior has been given by SAFRONOV (1969).

Calculations based on these formulae assume radial symmetry in the temperature profile. Initial inhomogeneities result from the impact of large bodies. These were not scattered about a large area but were mixed with the material in a vôlume exceeding their own by roughly ten times. The craters became buried unter other layers formed by infalling material. So primordial inhomogeneities in temperature were produced. Such regions were areas where partial melting and therefore differentiation occured first. These primary inhomogeneities should be superimposed upon the radial symmetric temperature distribution.

The heating mechanism by dynamo induction from intense solar wind emitted by the early Sun passing the stage of a T-Tauri star, as it has been proposed by SONETT, is rather doubtful (LEVIN, 1972). It shall not be discussed here.

From equations (4.1) and (4.2) under consideration of the very effective heat transfer by convection, what is taken into account by introducing a modified thermal diffusivity $k_V = Nuk_C$, with Nu — Nusselt number, k_C — thermal diffusivity and $k = k_C + k_V + k_I$, k_I — "eddy diffusivity" connected with impact stirring, one reaches at the initial temperature profile.

The second feature, that influence the differentiation process, is the melting curve of the various mantle materials, Fe and FeS, especially of the eutectic,

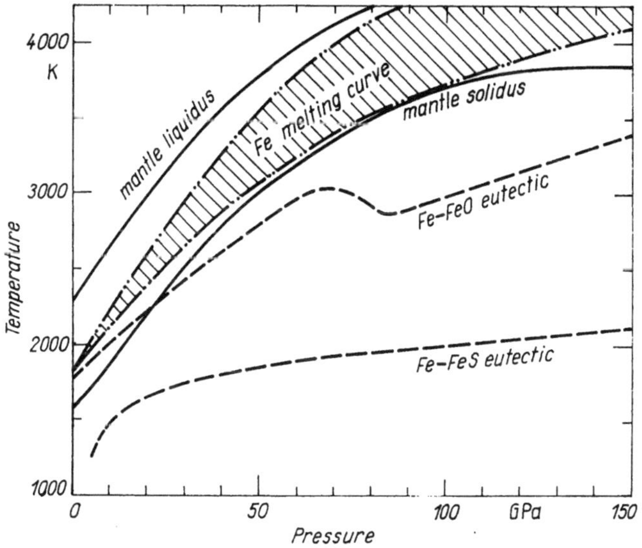

Fig. 4.9. Melting curves of several important materials (**from** Mc CAMMON et al., 1983).

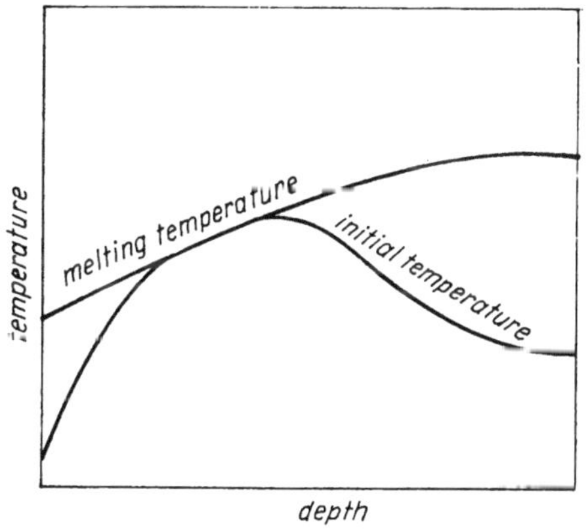

Fig. 4.10. Initial temperature distribution and melting curve of the Earth (according to VITJAZEV and MAJEVA, 1980)

and FeO. Figure 4.9 shows such phase diagram for these materials. The Fe—FeS eutectic is taken from USSELMAN (1975) and the melting interval of the mantle from KENNEDY and HIGGINS (1973).

Combining these results with the above considerations one arrives at Fig. 4.10. The temperature profile starts with a low value at the surface, increases towards the centre and passes through a maximum in the upper mantle. Then it decreases again and reaches a relative low temperature in the centre. That clearly indicates that differentiation commences in the upper mantle and proceeds into the centre. This process can have begun early in the history of the planet. Arguments for this may be found from the analysis of the initial state (see for example VITJAZEV and MAJEVA, 1980 or VITJAZEV, 1982).

Numerical results from CORADINI et al. (1983) clearly reveal, that core formation may start within the frame of accretional planetary growth if the

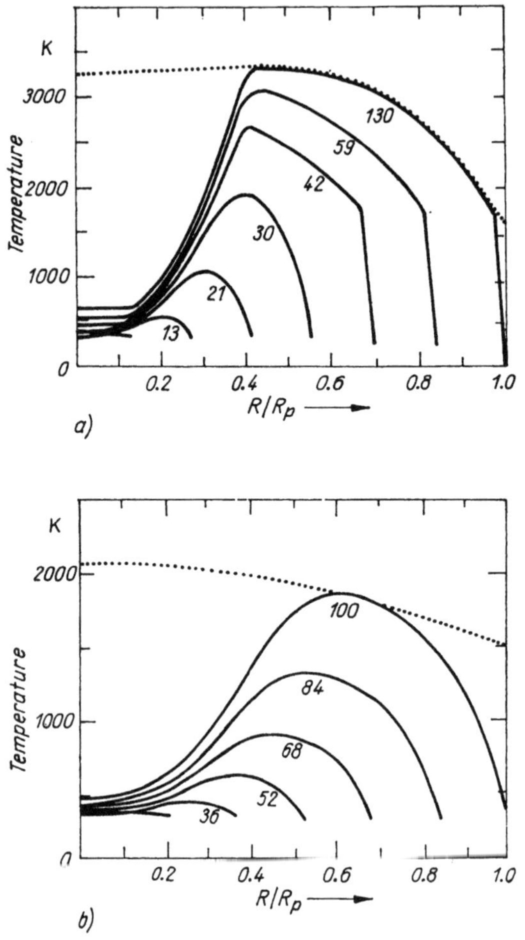

Fig. 4.11. Early thermal profiles for Earth and Mars (from CORADINI et al., 1983).
numbers: time in Ma
dotted: melting curve of silicates
a) Earth b) Mars

planet has grown to about 70% of its present radius (this is the value for the Earth and depends on the initial conditions). A large molten shell occurs at 42 Ma as indicated in Fig. 4.11a for the standard Earth model. This process is illustrated in Mc Cammon et al. (1983).

For Mars a molten shell occurs only at the end of the accretion process so that this planet has a slightly different evolution scheme than the Earth (see Fig. 4.11b). The thermal profiles and the size of the molten shell depend considerably on the initial conditions as impact energy, radioactive energy and the assumptions for the various physical parameters. So modifications are possible.

With these initial conditions the core formation process is considered.

Models for Core Formation

The core forming process is the most important evolutionary change in the structure of the planetary interior. It has consequences for the thermal state, the mechanical quantities, e.g. axis and period of rotation and for the convection in the mantle and therefore possibly for plate tectonics and other physical processes. Furthermore, it influences significantly magnetic and gravitational fields.

As has been discussed above, there is broad common sense today that the differentiation of core and mantle is based on the separation and redistribution of iron and silicates. The starting-point is a planet with homogeneous interior, that means that the two basic constituents are distributed evenly over the whole planet. This can be based on the assumption that the infalling matter contains both iron and silicates in solar abundances (cf. chapter 4.1.). Possibly existing inhomogeneities are of the order 10...100 km.

According to thermal evolution the temperature curve approaches the melting curve of iron and reaches it first in the upper mantle. This iron melting will occur before melting of the major silicate fraction, because any alloying constituent will depress the melting point (for instance sulphur, oxygen and, as discussed in the last time by Suzuki et al., 1983 also hydrogen (for this see also Fukai and Akimoto, 1983)). The liquid iron will penetrate through the silicate matrix, migrate downward and form the core.

Several different models have been developed for this formation process. According to Elsässer (1963) big iron or iron compound (as FeS) blobs with a radius of a few hundred kilometres will be formed in a molten region of the upper mantle. Due to a Rayleigh-Taylor instability these blobs will move slowly through the less dense mantle and accumulate in the area of the core (cf. Fig. 4.12).

Another model is that proposed by Vitjazev and Majeva (1977). In this model the molten iron or iron with alloying elements (FeS) of the upper mantle will collect in a spherical layer migrating slowly downward by melting the iron below the phase boundary. The silicates remain unmolten and form the mantle. The released silicate grains from the primordial core float up and penetrate through the iron layer to merge with the mantle above (cf. Fig. 4.13).

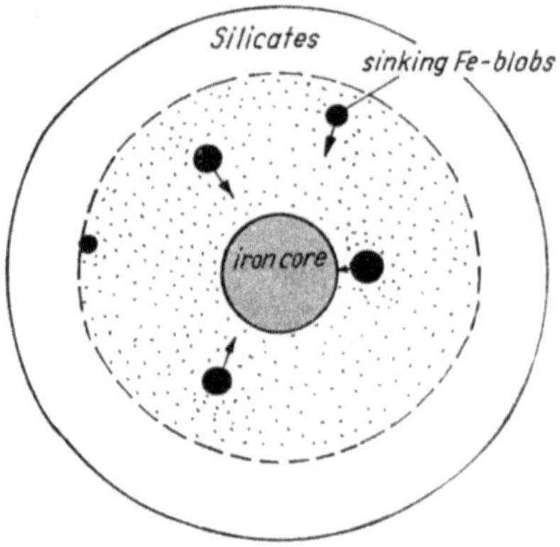

Fig. 4.12. Core formation process according to ELSÄSSER.

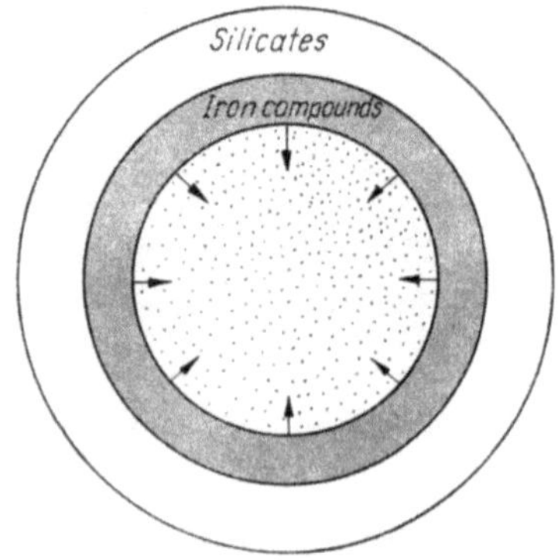

Fig. 4.13. Scenario for core formation by VITJAZEV and MAJEVA (1977).

The velocity of motion of a phase boundary between solid and liquid phases is mainly determined by the process of thermal diffusion. This very slow process continues for a long time, approximately 10^9 years. Taking into account, that due to the geomagnetic field the oldest magnetized rocks are about $3.5 \cdot 10^9$ years old one should prefer somewhat faster core formation models. This can be an indication that a faster growth of planetary bodies occured. Such a model of a faster core-formation is the scenario by STEVENSON (1981), the third model introduced here. The starting-point is similar to that of the model by VITJAZEV and MAJEVA. A spherical shell of molten material assembles in the mantle around the inner colder material (Fig. 4.14). This configuration is unstable and it may transform to a stable one by spontaneous symmetry breaking on a time scale of hours.

In this process the cold inner material is cracked and moved outward. That rapid event is followed by a steady-state process of formation and downward migration of iron blobs with typical sizes of 1...10 km.

Other models for the core formation process should be mentioned only in short: the models by ARTJUSHKOV (1970) or KEONDJAN and MONIN (1977) start with the assumption, that the differentiation begins in the centre of the planet with the separation of iron and silicates. The problem is to find such a separation process that works at the relatively low temperatures in the centre. KEONDJAN and MONIN proposed therefore a surface reaction for the separation of iron from a mixture of silicates.

But up to the present time there is no possibility of deciding between these two different approaches or between the three main scenarios mentioned above.

The time scale for core formation should also be discussed in relation to the problem of meteorites. As has been explained in detail in part 3 of this book, there is a great variety of meteorites (iron, stone, stone-iron) that must result from a highly differentiated parent body. Because the age of the meteorites

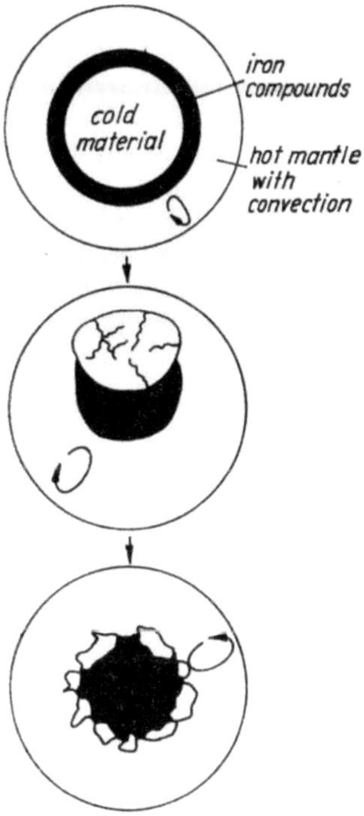

Fig. 4.14. Early stages of the core formation by spontaneous symmetry breaking according to STEVENSON (1981).

is in the order of $4.55 \cdot 10^9$ a it can be concluded that the differentiation of the parent body must have been a very quick process ($< 10^8$ a). In this way quick scenarios of accretion with early partial melting and early core formation seem to be favoured. An alternative approach would be that of a "fast" protoplanetary origin of planets.

Energetics of Core Formation

The energetics of core formation is treated here for the Earth as an example; it may easily be adopted for the other terrestrial planets as well due to their similar structure and differentiation scheme.

Assuming a primary undifferentiated homogeneous Earth, the core formation with its large-scale fractionation into iron or iron compounds and silicates is accompanied by the release of a certain amount of gravitational energy. Estimations of this energy are based on several different models with variable specifications. Basically this is the difference of the internal energy between early homogeneous state and the later differentiated, but taking into account all acting influences is very difficult and therefore several assumptions and simplifications have to be chosen and included into the calculations. But conclusions from all these models reach at nearly the same point: a considerable rise in the mean temperature. For example FLASAR and BIRCH (1973) calculated the released energy to be about $2.5 \cdot 10^6$ Jkg^{-1} while TOZER (1965) arrived at $2.0 \cdot 10^6$ Jkg^{-1} for the energy release (TOZER, 1978 estimated $1.7 \cdot 10^6$ Jkg^{-1} for the Earth). This heat and therefore the rising temperature influences strongly all processes and physical parameters in the Earth's interior such as thermodynamic quantities (compressibility, thermal conductivity, viscosity) and significantly convective processes. Most of this energy is converted into heat as many different investigations show, taking into account rotational and strain energy, too. So the energy mentioned above would be sufficient to heat the Earth by about 2000 K (FLASAR and BIRCH, 1973) while for instance SHAW (1978) estimated a rise of nearly 3000 K. One open question remains unsolved: how is the energy distributed in the Earth's interior — homogeneously or heterogeneously?

To some extend this question depends on the preferred model. A simple model was developed by VITJAZEV (1973). He calculated the radial distribution of the released energy under several restricting assumptions:

(i) incompressibility of the Earth
(ii) no change of the gravitational acceleration g due to the proceeding of the core forming process
(iii) constant density difference $\Delta\varrho$ between iron and silicates, no equation of state is taken into account
(iv) the question of heat transfer between iron and silicates remains unsolved.

This model can be used to get first rough estimations for the released heat. Within this model the Earth is described to be composed initially of two different components (iron and silicates) with densities ϱ_1 and ϱ_2. The inclusions are spread homogeneously over the volume with a volume fraction c independent of depth. Sinking of such inclusion with radius a along a path element δr releases potential energy

$$\frac{4\pi}{3} a^3 \Delta\varrho g \delta r . \qquad (4.5)$$

For a shell with $R_\oplus \geqq r_i \geqq R_c$, $R_\oplus$-radius of the Earth and R_c-core radius, this yields a density of energy release per unit volume from sinking denser iron

$$\varepsilon(r_i) = \frac{\frac{4\pi}{3}(R_\oplus{}^3 - r_i{}^3)\, cg\, \Delta\varrho\delta r_i}{4\pi r_i{}^2 \delta r_i} = \frac{(R_\oplus{}^3 - r_i{}^3)\, cg\, \Delta\varrho}{3r_i{}^2} . \qquad (4.6)$$

In the shell $0 \leqq r_j \leqq R_c$ the less denser mantle material rises and

$$\varepsilon(r_j) = \frac{r_j(1 - c)\, g\, \Delta\varrho}{3} . \qquad (4.7)$$

The parameter c may be estimated from the condition $\varepsilon(r_i) \simeq \varepsilon(r_j)$ for $r_i = r_j = R_c$. This results in $c = R_c{}^3/R_\oplus{}^3$. Figure 4.15 illustrates the energy

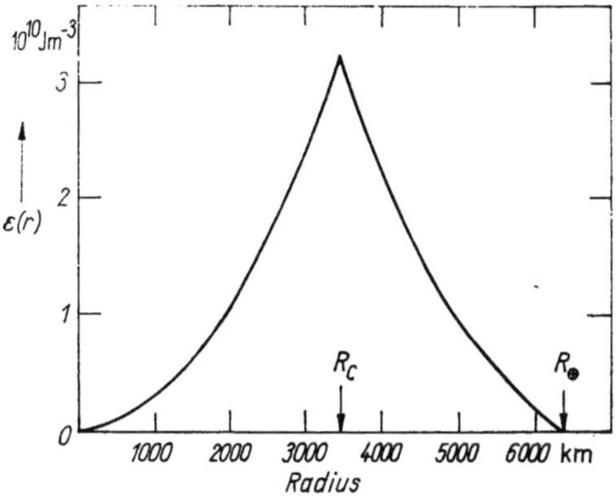

Fig. 4.15. Distribution of the energy density in the Earth's interior according to VITJAZEV (1973).

distribution. The maximum energy release occurs at the core-mantle boundary and the energy is distributed in nearly the same ratio as that of the volumetric abundances of core and mantle.

The total amount of released energy is obtained by integrating equations (4.6) and (4.7) for mantle and core respectively. The acceleration of gravity can be described to vary linearly through the core, reaching its surface value at the core-mantle boundary and therefore being constant through the whole mantle. With the appropriate radii of core and mantle it follows $c = 0.17$ and with $\Delta\varrho = 3.4\ \mathrm{gcm^{-3}}$ the total energy can be estimated: $8.7 \cdot 10^{30}$ J in the mantle and $3.5 \cdot 10^{30}$ J in the core. Summing up this corresponds to $2.0 \cdot 10^6\ \mathrm{Jkg^{-1}}$ what is in agreement with the values given above. This heat is not fully stored in the interior because the steep gradient of $\varepsilon(r)$ in the mantle will cause vigorous convection and the heat will be transported outward.

Of course, this model needs further improvements to bring it nearer to reality. For instance a secular evolution could be included taking into account the variations of the temperature profile with time (cf. CORADINI et al., 1983). Therefore only a certain shell of molten material exists (where the temperature curve reaches the melting curve of iron or iron compounds as FeS) where differentiation and energy release take place. In these time intervals only a part of the above estimated energy is released and so it is possible to establish an adiabatic temperature profile by convection throughout the whole Earth's evolution. The thermal processes will be discussed in the following section.

Improved models for core formation and especially the related energetics have been worked out for instance by SHAW (1978). The other terrestrial planets are treated in SHAW (1979). He concludes, that the core receives a substancial amount of energy (about $4.8 \cdot 10^6\ \mathrm{Jkg^{-1}}$) while the averaged value for the whole Earth is nearly $3.5 \cdot 10^6\ \mathrm{Jkg^{-1}}$. So the core with one third of the total Earth's mass gets nearly half of the available energy.

Consequences are that the energy release does not result in melting of the whole mantle as usually in the other models. MURTHY and HALL (1972) arrive at a similar conclusion. Their energy release of about $10^6\ \mathrm{Jkg^{-1}}$ would result in a temperature increase of 1000 K. In their model this value is an upper limit and so large-scale melting will not occur. On the other hand BIRCH (1965) discusses energies of about $2.5 \cdot 10^6$ Jkg^{-1} what is enough to raise the temperature by 2000 K. The occurance of whole mantle melting will accelerate this process, so that it could elapse in a geologically short time.

The steep temperature gradient leads to vigorous whole-mantle convection and effective mixing of the mantle materials may occur. Another feature is the enrichment of the core with the high pressure modification of potassium (MURTHY and HALL, 1972; BUKOWINSKI, 1976).

A rough estimation in a straight-forward analytical manner for all terrestrial planets is given by HOREDT (1980a and 1980b). He calculates the mean temperature rise of about 3300 K for the Earth from the difference of the gravitational and internal energies before and after core formation.

Another simple model for the differentiation and core forming process has been developed by DAVIES (1980). This model also assumes a homogeneous Earth as starting-point but it also includes convection by parameterized convection approximation (SHARPE and PELTIER, 1978). It results in mantle temperatures about 30% higher than at present.

All of these models depend significantly on the proposed time scales and the assumed initial temperature profile. This is dependent on the appropriate accretion scheme and the time of formation of the planet. The time of core formation strongly influences the temperature of the planetary interior immediately following the core formation and therefore the further thermal history and energy balance.

As already mentioned the considerations may be extended in a similar way to the other planets. A thermodynamic calculation of the thermal history of the terrestrial planets including the core forming process is found for example from TOKSÖZ et al. (1978).

After formation of the core, consisting of molten iron, cooling and solidification start. These processes are discussed now as further mechanisms for energy release. For describing the chemical differentiation it is first necessary to regard the phase diagram of iron sulphide. As usually accepted, the core does not consist of pure iron but of iron and some alloying elements as for instance sulphur as suggested by MURTHY and HALL (1972) or BRETT (1976).

Following VERHOOGEN (1973) the phase diagram of the Fe—S-system at high pressures should be similar to that of the Sb—S-system at normal pressure. With these assumptions the diagram of Fig. 4.16 has been constructed. The interesting feature of this is the occurrence of a liquid immiscibility field between the Fe-endmember and the eutectic. Above the liquidus curve there exist two liquids: a "heavy" liquid with low sulphur content (2.6 wt. %) and a "lighter" liquid with about 15 wt. % S. As has been pointed out by USSELMAN (1975) a small admixture of nickel would have an insignificant influence on the melting behaviour of the Fe—S-system; so the model could also be applied to a core with about 5 wt. % of alloying Ni. The model should be applicable too, if other possible light admixtures (as silicon or oxygen) in the core should exhibit a similar mixing behaviour.

Now the model of ascending droplets developed by FRANCK (1982b) is

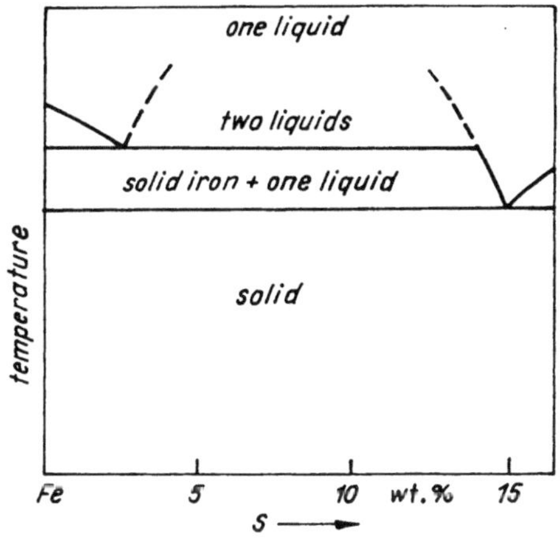

Fig. 4.16. Iron-rich part of the phase diagram of the Fe—S-system at high pressures.

introduced. It assumes that more than $4 \cdot 10^9$ years ago the whole Earth's core was in a liquid state. A slow cooling of the Earth led to freezing of the core material. As discussed by Loper and Roberts (1981) or Fearn and Loper (1981) this process started in the Earth's centre. The core material is an alloy with a small sulphur fraction. If a solid grows by freezing of such material the solid phase will contain a smaller mass fraction of the light material than the liquid. With Fig. 4.16 it easily can be concluded that the inner core consists of pure metal because it freezes from the heavy liquid with about 2.6 wt. % sulphur. This leads to a sulphur excess at the inner-core boundary. Because of the liquid immiscibility above the inner core this sulphur may not alloy with the remaining ("heavy") liquid. The only possibility is the nucleation of liquid droplets with about 15 wt. % S.

Permanent flux of sulphur causes the droplets to grow up to such a size where buoyancy is strong enough to overcome the drag forces. This radius may be estimated from similar considerations Stevenson and Salpeter (1977) used for their investigations of the helium rain in giant planets. According to Landau and Lifshits (1978) the relationship between drag force and gravitational force becomes

$$\frac{4\pi}{3} (\varrho - \varrho') R^3 g \geqq \frac{\pi}{2} C \varrho u^2 R \tag{4.8}$$

with ϱ — density of surrounding "heavy" liquid; ϱ' — density of droplets ("light" liquid); u — characteristic velocity of ascending droplets; C — drag coefficient. The densities are found by linear extrapolation from Fig. 5 of Ahrens (1979). The characteristic velocity is adopted from Stix (1977) as well as the kinematic viscosity for estimating the Reynolds' number and therefore C. Applying Stokes' formula to C (Landau and Lifshits, 1978) the radius is about $R \geqq 6 \cdot 10^{-5}$ m.

The stability condition for these droplets is found to be

$$R \ll \left[\frac{\alpha}{\varrho g}\right]^{1/2} \tag{4.9}$$

(LANDAU and LIFSHITS, 1978), with α — surface tension of the droplet. Appropriate values for α are obtained from experimental data for the cristallization of iron from its melt. Following LEWIS (1975) the ratio of surface energy $4\pi R^2\alpha$ and latent heat L is about 0.45. With values for L it follows the stability condition $R \ll 1$ m. The characteristic growing time of such small droplets is determined by the diffusion of sulphur atoms in liquid iron. With the material diffusion coefficient D_S (FRENKEL, 1957a) the time τ is simply related by

$$\tau = \frac{R^2}{D_S}. \tag{4.10}$$

This time τ is about 1 s what is several orders of magnitude less than typical ascending times in the outer core. These are in the range of $10^9 \ldots 10^{10}$ s. So the growing process is very fast compared to ascent.

The droplets ascend as for simplicity assumed from the inner core boundary. In this process the small drops accumulate to form greater clusters. This follows from some assumptions of dynamo theory. The size of these clusters is about several kilometres. These assemble near the core-mantle interface in a

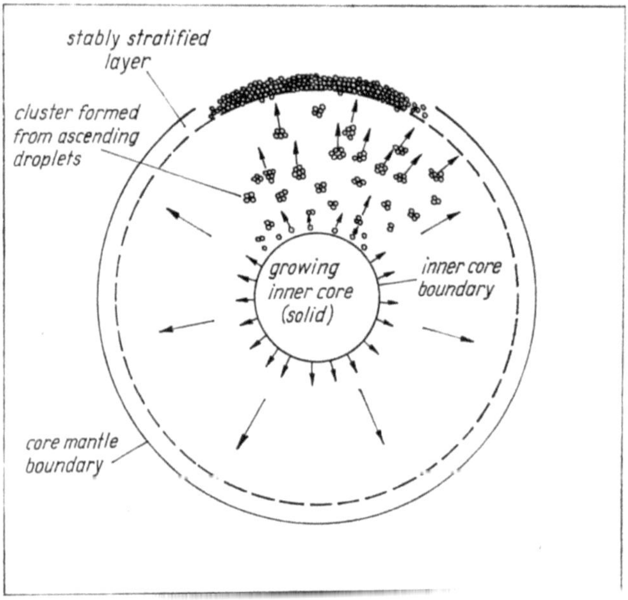

Fig. 4.17. Inner-core growth by ascending droplets (influence of rotation on the droplets' motion not taken into account).

possibly stably stratified layer. Figure 4.17 illustrates this. The present inner core radius and the densities of "heavy" and "light" liquids result in a layer thickness of about 11 km with possible lateral inhomogeneities in the range of the size of the formed greater clusters. As it is seen the transport of light material to the upper core boundary leads to a rearrangement of the core matter accompanied by a release of potential energy.

This energy is estimated by Franck (1982b). With $J_T(r)$ — the number of droplets that nucleate per unit time it can be calculated from the integral (clustering has no influence)

$$P = \int_{R_{ic}}^{r_e} \frac{4\pi}{3} (\varrho - \varrho') R^3 g(r) J_T(r) \, dr . \tag{4.11}$$

The radius r_e is that, where ascending ends — for simplicity it is assumed to be $r_e = R_c$. The unknown quantity $J_T(r)$ can be related to the flux of light material $I(r)$ from the inner-core boundary (Loper and Roberts, 1981) and equation (4.11) becomes

$$P = 6.6 \int_{R_{ic}}^{R_c} \left[\frac{\varrho - \varrho'}{\varrho'}\right] g(r) \, I(r) \, dr . \tag{4.12}$$

The quantity $[(\varrho - \varrho')/\varrho']$ is approximated by 0.13 for the whole outer core and equation (4.12) is evaluated in the same way as by Loper and Roberts (1981). The result is

$$P = \left(\frac{0.86}{\bar{\alpha}}\right) \cdot 6.3 \cdot 10^{11} \text{ W}, \tag{4.13}$$

where $\bar{\alpha}$ is a measure for volume change in the Fe—S-system according to changes in the sulphur content. Loper (1978) estimated $\bar{\alpha} = 2.4$. Then from equation (4.13) a power $P = 2.26 \cdot 10^{11}$ W follows. The dissipation due to diffusion can be neglected considering that the diffusion coefficient from Einstein's relation for the droplets is about six orders of magnitude smaller than the material diffusion coefficient. Other relevant parameters (volume change due to change of composition, mixing heat, etc.) alter less than the diffusion coefficient. That is why the full power is available to drive the Earth's dynamo because all suppositions for a dynamo action are present: the motion of an electrically conductive material in a rotating system. As Stiller et al. (1975) described, the conductivity of FeS changes in the pressure range of about 1 GPa from that of a low gap semiconductor to a metallic one.

Loper and Roberts (1981) used a formula to relate the Ohmic power dissipation P_0 to the average toroidal magnetic field B_a

$$P_0 = 10^{15} \frac{\text{W}}{\text{T}^2} [B_a]^2 , \tag{4.14}$$

what in turn is only a rough estimation because of the uncertainties in the

Ohmic dissipation associated with a given field and the relation between average and maximum toroidal field. If P_0 is the power resulting from the model then an average field of $1.5 \cdot 10^{-2}$ T or 150 G is obtained. This could be classified as one of mean size.

In addition it should be mentioned that in the model described here the droplets ascend from the inner-core boundary to almost the core-mantle interface. If one assumes only a small liquid immiscibility just above the liquidus curve then the droplets would rise only to a certain radius and vanish somewhere in the outer core because of full miscibility in that area. This would cause chemical convection in the remaining core. So two different mechanisms could act simultaneously.

4.2.3. Thermal Evolution of the Terrestrial Planets (U. Schmit)

There is a tight but rather complex connection between the accretion history of the planets at the one side and the thermal evolution of the planetary bodies at the other side. Especially accretion implies the release of a huge amount of gravitational energy, converted into kinetic energy of impact. Without heat transfer a temperature rise of roughly

$$c_p \Delta T = \frac{\gamma M(R)}{R} \tag{4.15}$$

would follow. Here c_p is the specific heat, ΔT the increase of temperature, γ is the gravitational constant, R the radius of the growing planetary embryo and $M(R)$ is its mass. In the Earth temperatures would reach 50000 K at the end of accretion, cf. Hanks and Anderson (1969). Indeed temperatures of this level will never be reached, because most of the heat is transported outward and radiated away. More detailed calculations are necessary.

Connection of thermal evolution and accretion history is of interest for both sides. A thermal history calculation is mathematically spoken an initial-boundary-value problem. Assuming a "start value" and boundary conditions makes it possible comput the thermal evolution of a planet during the 4.5×10^9 years of its life. The "start value" is proposed to be taken from accretion scenarios. On the other side each accretion theory should be able to pass a test according to its thermal consequences in the light of thermal history models, of geophysical and geological facts.

The present paragraph is an attempt to enlight some connections between both areas; there is no claim for completeness, of course.

Hanks and Anderson (1969) proceeded model calculations for the early thermal history of Earth and Mars. They assumed, that the Earth accumulated by a gravitational collaps from dust particles in a gas-free environment. Neglecting heat transport inside the growing body, the temperature at the surface is determined by the balance between gravitational energy release during impacts and radiation from the surface into the surrounding space:

$$\varrho \frac{\gamma M(R)}{R} \frac{\mathrm{d}R}{\mathrm{d}t} = \varepsilon\sigma\left(T^4(R) - T_b^4\right) \tag{4.16}$$

where t — time during the accretion process, σ — Stefan-Boltzmann constant, T_b — temperature of the surrounding, i.e. the equilibrium temperature of particles in the surrounding space, for which 278 K was taken, ε — emissivity parameter, ε is of the order of one, for blackbody radiation $\varepsilon = 1$ would be true. $T(R)$ is the surface temperature of the growing embryo. If dR/dt is known, integration of (4.16) over the time yields an initial temperature $T_0(r) \equiv T(t_0, r)$ inside the whole planet. t_0 denotes the end of accretion and the beginning of the subsequent thermal evolution of the planet; r — radial coordinate.

Because of the lack of better knowledge HANKS and ANDERSON used for the accretion rate two plausible forms

$$\frac{dR}{dt} = A \sin \alpha t, \tag{4.17}$$

$$\frac{dR}{dt} = Ct^2 \sin \gamma t. \tag{4.18}$$

The constants were chosen to give some special total accretion times t_{acc}. HANKS and ANDERSON calculated several models including various accretion times. For $t_{acc} \sim 10^6 - 10^8$ years the Earth remains too cool to allow core formation, in their opinion. Only for

$$t_{acc} \leqq 10^5 \text{ years} \tag{4.19}$$

they got satisfactory results, i.e. core formation.

Note by the way, that in both formulas (4.17) and (4.18)

$$\frac{dR}{dt} \to 0 \quad \text{for} \quad t \to t_0, \tag{4.20}$$

i.e. the accretion rate vanishes at the end of accretion, because of increasing lack of material that could be accreted at the end. It follows, that the surface temperature for the newly completed planet is always low:

$$T(t_0, R) \approx T_b, \tag{4.21}$$

independent of the total accretion time, as HANKS and ANDERSON state in all of their models.

Another early work is the paper of TOKSÖZ et al. (1972) on the thermal evolution of the Moon. For the initial temperature they used a similar formula like (4.16):

$$\varrho \frac{\gamma M(R)}{R} \frac{dR}{dt} = \varepsilon\tau(T^4 - T_b^4) + \varrho c_p(T - T_b) \frac{dR}{dt}. \tag{4.22}$$

The last term in (4.22) admits the fact, that the cool accreting matter heats itself too, before it can heat the rest of the body. (4.22) results in a quite similar initial temperature curve as HANKS and ANDERSON (1969) got for the Earth. TOKSÖZ et al. used the accretion-rate law (4.18); their total accretion rate t_{acc} results from considerations about the Moon's surface. The Moon has an accentuated crust that must have differentiated out of mantle at some time. An early heating including partial or complete melting seems unavoidable. In the above scenario an early heating demands a short accretion time. Toksöz et al. conclude: "One can estimate the duration of accretion by matching this model to the temperatures required to cause melting to a specified depth. To differentiate an anorthosite crust about 10 km thick requires melting to a depth of about 500 km. This can be achieved, if the total accretion time is about

$$t_{acc} \sim 100 \text{ years}. \tag{4.23}$$

Total accretion time may be made arbitrarily long, however, by appropriately lowering the assumed emissivity of the lunar surface."

Further details of the calculations of Toksöz et al. will be discussed below.

According to the above estimates (4.19) and (4.23) it is to remarke that reality is more complex, because the connection between the accretion time and early heating depends drastically on the scenario chosen. In this paragraph we shall confine to the terrestrial planets, including the Moon, since the knowledge of thermal properties of the outer planets is rather poor, so they are not suitable objects for the aim of this paragraph.

The theories leading to (4.19) and (4.23) assume a direct gravitational collaps from dust to planets without intermediate stages of growths. This means, in principle, that the surface of the growing planet is beaten by a large amount of single dust particles. A small part of the heat of impact remains immediately below the surface, the greater part is radiated back into space.

A new aspect came into discussion by the work of Safronov (1969) and coworkers: In the first step the dust coagulates somehow into bodies of kilometer-size, the so called "Planetesimals", while in the second step these planetesimals accumulate by mutual collisions into a planet. Different from the former scenario, now bodies of km-size hit the surface. A certain part of the impact energy is buried as heat deep enough to be restored in the planetary body during the whole accretion time and partly until the present time. Independent of t_{acc} there seems always to be enough energy for an appreciable heating of at least the outermost few hundred kilometers of the terrestrial planets. Now we shall discuss that point in more detail. First, Safronov (1978) states: "Attempts by some authors to obtain initially hot planets by assuming a short time scale for their formation are unsuccessful because this assumption is not supported by the study of the planetary accumulation process. There is another factor which can lead in principle to high temperatures of the outer parts of the planets — the impacts of large bodies".

Further he concludes: "The result depends substantially on the sizes of impacting bodies but almost not at all on the time scale of accumulation."

Kaula (1979) reaches at similar conclusions: "These requirements for early heat have often been misinterpreted as requirements for rapid accretion, of the order of 10^3-10^4 years accretion time (Cameron, 1973; Ringwood, 1975). This need for rapid accretion is based on the dubious assumption that the accreted material was all in very small particles, such that all the heat is deposited in a thin layer so that it can rapidly radiate away (Hanks and

ANDERSON, 1969; MIZUTANI et al., 1972). Further KAULA concludes: "There is a slight *positive* correlation of heating with growth time, since a longer time implies larger planetesimals."

The heating of a growing planetary embryo was modeled by SAFRONOV (1978), KAULA (1979) and CORADINI et al. (1983). During the impact of a planetesimal a fraction h of its kinetic energy is buried below the surface via shock waves, and is going into heat at last. For the terrestrial planets h is of order 0.1—0.4, cf. CORADINI et al. (1983).

The impact energy is just the gravitational energy released, as it was already expressed in formula (4.15). Therefore, neglecting internal heat transport for the moment, the temperature in the planetary embryo is given roughly by

$$T(r) = \frac{h}{c_p}\left(\frac{\gamma M(r)}{r} + v_\infty^2\right) + T_b; \tag{4.24}$$

$M(r)$ is the mass of the sphere with radius r inside the growing body, v_∞ is the velocity of the infalling planetesimal "at infinity". Because

$$M(r) = \varrho \frac{4}{3} \pi r^3$$

(ϱ — density of matter) the temperature at the centre is very low and increases roughly as

$$T(r) \propto r^2 \tag{4.25}$$

cf. STEVENSON (1981), KAULA (1980).

Equ. (4.25) holds in the main part of the interior, while the outermost layers are governed by the boundary condition, i.e. the temperature at the surface. In the frame of SAFRONOV's scenario planetesimals hit the surface, causing local melt in an area of a few kilometres. A crude estimate of the time necessary to decrease the temperature in the heated area by radiation down again onto the mean surface temperature $T_b \approx 300$ K is

$$\tau = \frac{H_m c_p \varrho \, \Delta T}{\sigma(T^4 - T_b{}^4)} \tag{4.26}$$

where H_m is the thickness of melted material. For a typical planetesimal impact τ is less than a year, i.e. a year after the impact the surface is cold again.

On the other hand the time τ_i between two subsequent impacts of kilometre — sized planetesimals into the same area is about 10^5 years. τ_i is very much greater than τ, i.e. occurs no *global* heating of the surface by impacts. At a given time there are only a few single short-lived heated regions. Nevertheless every place at the surface is melted repeatedly by subsequent impacts. So it is correct to say that the whole surface was molten, but not all parts at the same time.

Exactly spoken the mean surface temperatur $T(R)$ is a little bit higher than the temperature T_b of the surrounding space, to radiate away the overall heat released during impacts via the

$$\varepsilon\sigma\left(T^4(R) - T_b{}^4\right)$$

term, but because of the long accretion time of $\sim 10^8$ years the difference is as small as

$$T(R) - T_b = 6\,\mathrm{K}, \tag{4.27}$$

cf. KAULA (1979). Therefore the surface temperature can be considered as a fixed point, it corresponds with the temperature of the surrounding space.

In early phases of evolution the growing planet exhibits a temperature distribution whose principle features are shown in Fig. 4.18: Low central temperature, an increase of $T(r) \propto r^2$, a maximum somewhere below the surface and a strong drop of the curve to reach the surface temperature $T(R) \approx T_b$. In the outermost layer heat is conducted along the temperature gradient.

The picture changes, when $T_{\max}$ approaches the melting regime since convection occurs. As soon as the viscosity is small enough convective motions develop at first in a region around $T_{\max}$, progressively going inward. It is well known, that the temperature distribution $T(r)$ follows approximately an adiabat:

$$\frac{\partial T}{\partial r} \approx \frac{\partial T_{\mathrm{ad}}}{\partial r}. \tag{4.28}$$

The adiabatic gradient is a given parameter of matter as well as the melting curve $T_m(r)$. For planetary matter it has been shown, that

$$\left|\frac{\partial T_{\mathrm{ad}}}{\partial r}\right| < \left|\frac{\partial T_m}{\partial r}\right|, \tag{4.29}$$

.e. the melting curve is steeper than the adiabat. With the onset of melting viscosity decreases sharply, so convection becomes efficient enough to drive away all the heat and prevent a further increase of T. Therefore $T_m(r)$ is an upper boundary for the whole planet:

$$T(r) \leqq T_m(r) \quad \text{for} \quad 0 \leqq r \leqq R. \tag{4.30}$$

Clearly T approachs T_m first at the radius r_1 corresponding to the maximum of the temperature curve, cf. Fig. 4.18.

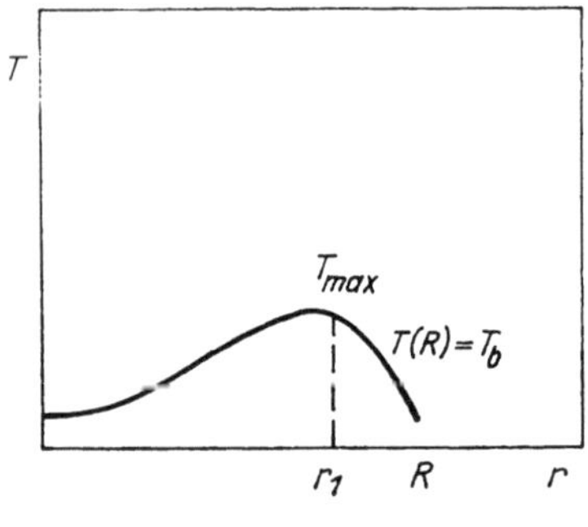

Fig. 4.18. Temperature distribution in a growing planetary "embryo".

The subsequent evolution proceeds as follows: Despite increasing $R(t)$, for the outermost layers the picture remains the same, i.e.

$$T(R) = T_b$$

and $T(r)$ rises strongly with depth, reaching T_m at some radius r_1, cf. Fig. (4.19). From r_1 into the deeper interior $T(r)$ follows closely the adiabat. Exactly spoken the temperature lies somewhere between the adiabat and the melting curve, because there are two opposite effects. At first without convection viscosity demands the temperature to follow the adiabat. But at low temperatures high viscosity can prevent convection. The critical temperature T_{cr}, allowing convection depends on the melting temperature T_m, it is

$$T_{cr}(r) < T_m(r) \quad \text{and} \quad T_m(r) - T_{cr}(r) \sim \text{const}:$$

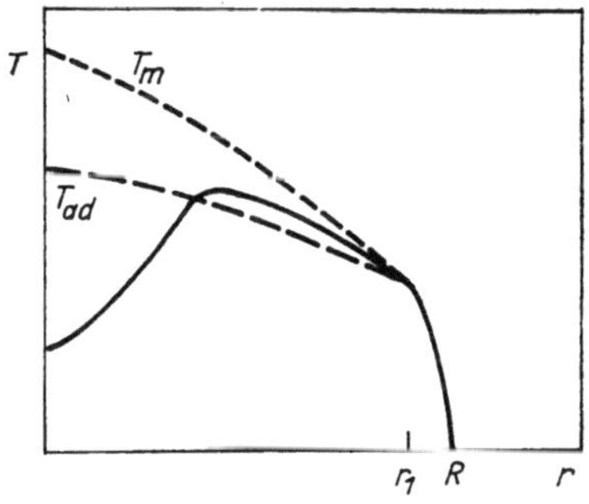

Fig. 4.19. Thermal state of planet Earth at the end of accretion, negleiting are formation.

The above formula tries to force $T(r)$ on a line parallel to $T_m(r)$. Reality makes a compromise, so $T(r)$ lies between $T_{ad}(r)$ and $T_m(r)$.

It may be wortwhile to remark here, that the notion "melting curve" for silicate matter is strictly speaking not correct. Composite matter has no sharp defined melting temperature but rather a more or less broad melting regime between a solidus temperature and a liquidus temperature. Below the solidus all matter is solid and above the liquidus all matter is liquid. In between it is in a state called "partially molten". In this paragraph the motion "melting curve" is a simplification to present the melting regime suitably.

The central part of the planet remains rather cold until core formation sets in, cf. Coradini et al. (1983).

Core formation is probably the most important event, besides impact energy, in the thermal history of accretion. There is a multitude of theories modeling that process. Because implications for $T(r)$ do not depend on details we sketch core formation like Stevenson (1981) did.

In this almost steady-state process, incoming iron-bearing planetesimals disseminate upon impact and Rayleigh-Taylor instabilities of the outer boundary layer occur. As this material migrates into hotter, subsurface regions, the iron alloy is mobilized and can percolate to form diapirs. The typical size of such iron blobs in this scenario is ~ 1 km, possibly more. Their downwards Stokes velocity depends on the viscosity of the surrounding mantle material. In a steady-state scenario, this is determined by the requirement that the excess heat of core formation is transported out by thermal convection.

Core formation is a "rapid" process, it keeps up with accretion. This is consistent with lead isotope data and with the evidence of a geomagnetic field in rocks as old as 3.5 billion years.

The hot iron blobs carry the heat into the deepest interior. This chemical differentiation is the only efficient mechanism of heat transport to the bottom, since thermal convection always drives the heat to the top. Independent of details the temperature curve after core formation is roughly the one, shown in Fig. 4.20: The situation in the mantle is mainly the same as in Fig. 4.19

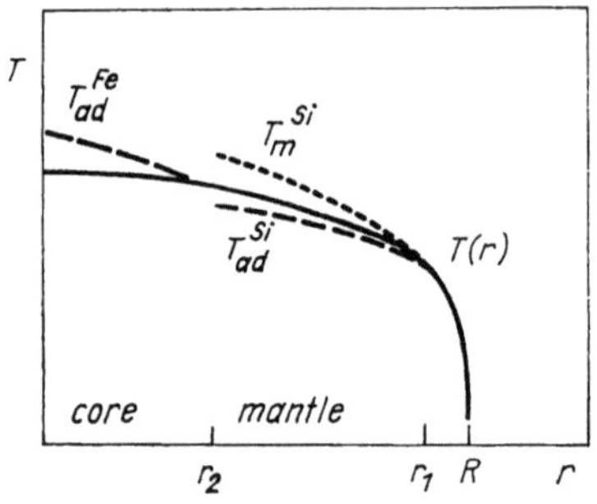

Fig. 4.20. Thermal effects of core formation.

but the core is rather hot. An important feature of $T(r)$ in the core is, that the temperature gradient is smaller than the adiabatic gradient, i.e.

$$\left|\frac{\partial T}{\partial r}\right| < \left|\frac{\partial T_{ad}^{Fe}}{\partial r}\right|. \tag{4.31}$$

T_{ad}^{Fe} denotes the core adiabat, while T_{ad}^{Si} is the mantle adiabat. (4.31) implies, that the core is "stably stratified" in thermodynamic sense, i.e. there occurs no thermal convection. Convective motions in spite of (4.31) are possible only if additional energy sources exist. This is the case, if there exists a solid inner core. Then cooling causes the solid inner core to grow, latent heat is released, in addition chemical differentiation at the inner-core boundary yields further energy, therefore in spite of (4.31) convection exists at least in the lower part of the outer core.

Because these motions in the fluid core generate the planetary magnetic fields our thermodynamic considerations suggest some constraints concerning planetary magnetism. Venus is quite similar to the Earth in size and composition, but it seems to have no dynamo. Indeed Venus rotates rather slow. However, even this slow rotation is fast enough to satisfy the demands of dynamo theory. It could be, that slight differences in composition, pressure and surface temperature relative to the Earth prevented inner core formation in Venus up to the present. Then the whole core was fluid throughout the last 4.5×10^9 years and because of (4.31) it was "stably stratified" from the beginning. Our conclusion is, that if it is correct, that the whole Venus core is still fluid, then it never had a dynamo in its history.

Here our opinion differs to models of STEVENSON et al. (1983), who got dynamo action in the past that ceased a billion years ago. The same holds for Mars. Mercury may have a mainly solid core with a small fluid shell at the top, where weak convection is located.

In the past, two major classes of planetary accretion theories were proposed "homogeneous" and "heterogeneous" ones. Homogeneous means, that all infalling matter has the same composition. Core formation is a secondary process, that takes place in the interior of the growing planetary embryo, cf. Fig. 4.19 and 4.20. Heterogeneousaccretion theories assume, that first an iron core arises followed by the formation of an overlaying silicate mantle. These hypotheses have the advantage that a core exists from the beginning and that there is no need to struggle with the uncertainties of core formation. Nevertheless they contradict to thermodynamics of planetary evolution. As discussed above differentiation into core and mantle is the only mechanism that can heat the deepest interior efficiently enough. Temperatures in the cores of the terrestrial planets amount to several 10^3 K. Models of "heterogeneous accretion" bear no possibility to explain the existence of these hot cores.

There are numerous calculation to explain further cooling of planets via convection, cf. LUBIMOVA (1982), SHARPE and PELTIER (1979), SCHUBERT et al. (1979), SPOHN (1984). A main result is, that planets behave as a "self-regulating system" caused by the temperature dependence of viscosity. The circle is the following: Greater heat generation implies a higher temperature; higher temperature means lower viscosity; a lower viscosity allowes for a faster convection but a faster convection reduces the temperature again. This is the reason why a certain temperature distribution cannot be exceeded, independent of heat generation by impacts. Accretion results in any case in an initial temperature as depictured in Fig. 4.20.

The further thermal evolution is a slow cooling with a decreasing cooling rate because viscosity grows continuously. Today the cooling rate is a few K per 10^8 years; the present day temperature distribution is rather independent ot the initial thermal state, cf. SPOHN (1984).

A major change in the view on the early thermal history of planets may arise by the blanketing effect of a hypothetical primordial atmosphere of solar composition. Such atmosphere hinders irradiation of heat into space producing high surface temperatures in this way. HAYASHI et al. (1980) calculated an Earth model with a surface temperature of 4000 K at the end of accretion. Of course a surface temperature of 4000 K would be quite another boundary condition than the usual 300 K of SAFRONOV. There exist no detailed investigations of this topic up to now. Certainly, the value 4000 K seems to be too high (it is above the boiling point of silicates!), but it should be much higher than 300 K and possibly it approaches the melting point of silicates. There is some support of HAYASHI's view, because the planets probably begin to accrete not in whole space but inside a thick disk of gas of solar composition

(H—He), cf. GOLDREICH and WARD (1973), WEIDENSCHILLING (1980), NAKAZAWA and NAKAGAWA (1981). During accretion it could be radiated away by solar wind; if not it must form a primordial atmosphere around the early planets. In any way the thermal consequence of a primoridal atmosphere is one of the most important future tasks in this field.

The last topic we have to deal with are planetary crusts. Does the surfaces of planets tell us anything about their origin?

The evolution of the Earth's crust is rather complex and still some what mysterious. It could have happened as follows, cf. SPOHN (1984): In the beginning the early Earth possesses no "crust" i.e. no well developed chemically distinct layer exists at the surface. Whole mantle convection takes place from the core boundary up to the surface. The whole mantle is in solid state everywhere, except in a small region 100—200 km below the surface, where partial melting occurs, cf. Fig. 4.20. Here chemical differentiation is initiated. In reaching the surface mantle material separates some lighter components to form a thin crust. During the first time convection is strong and pulls back into depth the newly formed crust by friction. Crust formation is delayed till 3.8 Million years before present, when convection becomes weaker and crustal plates are able to swim above the heavier mantle matter. This may be the reason why the oldest known rocks have an age of only 3.8 billion years, whereas the Earth age is 4.5 billion years. Since 3.8 billion years b.p. continental plates grow out of the mantle. Probably in the late Archaicum an equilibrium between crust and mantle turns up, so the volume of continental plates is constant since that time. Convection proceeds up to the present and continents drift on the mantle like floes on the water.

In 1915 ALFRED WEGENER wrote the book "Die Entstehung der Kontinente und Ozeane", it was the birth of plate tectonics. WEGENER proposed the existence of a single, connected supercontinent "Pangea" that was located mainly at the southern hemisphere. At 200 million years b.p. Pangea broke up and the pieces, i.e. the continents, drifted away to their present positions. But WEGENER's theory was poorly phenomenologic, i.e. he did not know the forces, that could moye such giant masses as continents. WEGENER for himself was an optimist; he thought that a new "Newton" would come and find the force.

Since the late fifties it became clear that convection yields the driving forces. But a few questions remained open. F.i. on the spherically symmetric Earth continents should exhibit some stochastic motion. Therefore it is difficult to imagine, how Pangea could form. Furthermore Pangea broke up 200 million years, b.p., this is only 1/20 of the Earth's history. What happened in the 19/20 before it?

This is a matter of future research; here we can only draw some tentative suggestions, partly in accordance with SPOHN (1984). A supercontinent as Pangea is not stable, because it is an isolating sheet, mantle heat is restrained below it and at last convection breaks the supercontinent; the pieces drift away. The mean drift velocity has a stochastic and probably a systematic component. Pangea was located mainly on the southern hemisphere. Continental motions show a systematic westward drift and especially a drift to the north (except the antarctic plate). Today the main part of continents is placed in the northern hemisphere; the remaining "southern continents" South America, Africa and Australia undergo the strongest northward motions. It may be that they drive to a new northern "Pangea". This supercontinent again will act as an isolating sheet, break up at any time and then convection pushes the continents to the south again. If this is correct continents undergo permanent oszillations between opposite Pangea-configurations. The dominance of the north-south direction may reflect some influence of rotation on convection.

Contrary to the dramatic features of the Earth crust history, Mercury and Mars show no evidence of plate tectonics, cf. Schubert (1979). Preliminary results of soviet radar investigations show no plate tectonics on Venus too, cf. Kuzmin (1985). These planets exhibit a one plate static surface. On the other side the general thermodynamic considerations made above demand convection in the interiors of all terrestrial planets. Furthermore there are some indirect indications of interior convection, cf. Schubert et al. (1979). (Certainly there exists interior convection, but it does not reach the surfaces.)

Our knowledge of the structure of Mercury, Venus and Mars is constrained by the lack of seismic data, especially the size of the core and the thickness of the crust are unknown (it is even possible, that there exists no well developed layer that we can call a "crust").

In any case there is convection in the interiors of the terrestrial planets, but convective motions dont' reach the surface. We have to expect that surface rocks are as old as the planets, i.e. 4.5×10^9 years. In reality the crusts underlie subsequent changes by vulcanism and a late meteorite bombardement.

Direct radioactive age determinations of surface rocks don't exist, indirect methods are inconclusive and partly contradictory. Mercury and Mars have similar surface topology as the Moon. It is possible that the age and history of the crust are the same (the Moon will be discussed below).

The soviet radar investigations yield a Venus morphology that suggests an age of only 1 billion years. This is surprizing and does not correspond with the general picture.

There exist several thermal evolution calculations of the terrestrial planets based on Safronov's planetesimal theory. For Earth and Venus there are no difficulties to satisfy observational contraints Safronov (1978), Kaula (1979), Coradini et al. (1983) and Arkani Hamed and Toksöz (1984). For the minor planets Mercury and Mars accretional heating and radioactive decay may also result in acceptable thermal history models, cf. Coradini et al. (1983).

The aim of this paragraph is to find some connections between planetary surface features and accretion theories via thermal history models. Unfortunately the problem can not be solved in a unique manner; there are at least 3 different accretion models with similar thermal implications, i.e.

a) accretion from dust in a short time (10^4 years),
b) accretion from planetesimals in a long time (10^8 years) and
c) accretion in a primordial atmosphere of solar composition (greenhouse effect).

The Moon was excluded from the above considerations on purpose. The Moon is an outsider in the solar system: despite of several lunar missions and much research on the topic its origin is still quite mysterious. At least the Moon does not fit into SAFRONOV's standard planetesimal theory. It may be that the Moon will play some key role in future investigations on terrestrial planets.

In a review of some of the main lunar features we follow TAYLOR (1979). Models for the internal structure of the Moon down to depths of 1000 km have reached a consensus. The lunar crust, which is less dense than, and chemically distinct from the interior, varies in thickness from 60 km to 100 km. The mantle exhibits some seismic discontinuities. The upper mantle or lithosphere extends from the base of the crust to 400—500 km; its seismic velocity gradients are negative. A minor discontinuity exists at a depth of 480 km, with a change to lower seismic wave velocities down to depths of about 1000 km. There is a major discontinuity at about 1100 km, with a very sharp decrease in wave velocities. The possible existence of a lunar core has been much debated but the evidence for it remains inconclusive. The seismic data restrict the core redius to 170—360 km.

Most of the surface is covered by so called lunar highlands. Age determination of surface rocks by lead isotope investigations show to peaks, one at 4.46 billion years and a second at 3.86 billion years b.p. The first is interpreted as the age of crust formation. Since the date for the condensation of the Solar System is usually given as 4.57 billion years b.p. accretion of the Moon followed by crust formation was accomplished in 10^8 years. The other peak at 3.86 billion years is often interpreted as a late heavy planetesimal bombardement.

The mare basalts, although prominent visually, cover only 17% of the lunar surface. Their mean depth is 1 km, therefore they constitute less than 1% of the crustal volume. Their concentration on the Earth-facing hemisphere still waits for convincing explanation. The mare basalts are derived by partial melting from depths of 100—400 km, hence they provide in principle information about mantle composition. Dated flows extend from 3.9 to 3.16 billion years b.p. When contemplating the whole spectrum of basalt types derivation from a cumulate source region remains the most promising model.

Closely related to the possible existence of a core is the lunar magnetic field. Lunar surface rocks possess a stable natural remanent magnetization (NMR), which indicate the former presence of a now vanished magnetic field. Many theories have been proposed for the origin of the field. They fall within the three categories:

1. the field is random due to localised near surface origin;
2. external uniform fields;
3. internal dipole field.

No decision seems possible at present between these alternatives; we have to wait for a future lunar orbiter.

NMR of lunar surface rocks exhibits some surprising features:

(A) the oldest samples (> 3.9 billion years b.p.) are among the most weekly magnetized;

(B) between 3.8 and 3.6 billion years b.p. a substantial lunar surface field existed, comparable in intensity to the Earth's magnetic field;
(C) later the field dropped to zero.

An explanation of such a field in the light of the above thermodynamic investigations could be a lunar core, first completely fluid and stably stratified. Between 3.8 and 3.6 billion years the core froze over, generating dynamo action. Since that time the core is in solid state and dynamo action ceased. A detailed quantitative dynamo model including energetic aspects is still a future task, cf. Cisowski et al. (1983).

At the other side, the apparent coincidence of three major events in lunar history, the end of the late heavy bombardement, the onset of mare volcanism, and the period of intense magnetization of lunar samples invite speculations on a common cause for these events. The least implausible explanation is, that the events are related to a close approach of the Moon to the Earth, as capture theories of lunar origin demand. Nevertheless the geomagnetic field alone cannot account for the necessary lunar surface fields.

There remains the possibility that a combination of both above mechanisms account for the magnetization of the lunar samples. It seems, that a former fluid core is needed, unless the complete recording of the magnetic field by a future orbiter favours a localized near surface origin.

In this paragraph we want to connect accretion theories with thermal history models and the present state of the Moon. Every accretion model has to provide at least two results, i.e.

(1) crystallization of a 60 km thick crust, chemically distinct from the mantle;
(2) formation of an iron core.

The early heating of the Moon by planetesimal impacts in the sense of Safronov was simulated numerically by Kaula (1979), using a total accretion time of 3×10^7 years. In contrast to the terrestrial planets the Moon remains too cold to allow silicate melting and probably the same holds for iron compounds. Therefore in Safronov's theory a binary accretion of Earth and Moon from planetesimals appears to be infeasible to explain the lunar crust.

Melting of the uppermost part of the lunar mantle and crust crystallization demands a shorter accretion time as proposed by Möhlmann (1985) and in earlier papers by Toksöz et al. (1972), Hanks and Anderson (1972), or an intense tidal heating during a possible capture of the Moon. Detailed thermal history calculations for the latter case is a further future task.

Little progress has been made in the field of accretion scenarios for the Moon. The three old hypotheses, i.e.

(1) binary accretion,
(2) fission from the Earth,
(3) capture of an already formed Moon by the Earth,

all survived up to the present. Nevertheless fission suffers from the old angular momentum problem; we believe that the formation scenarios of the Moon are

(I) some kind of binary accretion in a short time (possibly as short as 10^3-10^4 years)

or

(II) a separate formation of the Moon and major heating during a subsequent capture.

In any case the Moon is one of the most important objects of the Solar System. Lunar origin is an open question, it should initiate further investigations and future missions.

Conclusions

Concerning the origin of planets out of a preplanetary nebula mainly two theories were developed in the past decades, the "giant gaseous protoplanet" hypothesis of Cameron and the "planetesimal" theory of Safronov. The basing idea of the giant protoplanet scenario is the following.

Planet formation starts with a flat preplanetary nebula, an envelope of gas and dust, surrounding the young Sun. The nebula is roughly of solar composition, it consists to 99% of gas (H and He) and to 1% of the heavier elements (mainly in the state of solid grains, i.e. "dust").

Cameron suggested the nebula itself to be unstable against gravitational collaps, if the nebula's mass is comparable to the mass of the Sun. Gravitational instabilities lead to hot giant protoplanets of solar composition. Their further evolution is characterized by cooling and convective motions. Possibly the dusty component is able to establish grains of cm-size or more and these grains migrate into the centre of the protoplanet, to form a "rocky" core there. Later on the gaseous H—He envelope disappears completely (terrestrial planets) or partly (outer planets) by the action of the solar wind.

Safronov's planetesimal theory bases on a preplanetary nebula of low mass (i.e. only a few per cent of the solar mass). The low mass nebula is stable against gravitational collaps, but possibly the dust migrates into the median plane, establishing a very thin sheet of high local density. While the gaseous envelope remains stable against gravitational collaps there may arise instabilities in the dust sheet. In any way the dust coagulates into "planetesimals", i.e. stony bodies with diameters in a range, reacting from centimeters to kilometers and more. These bodies grow slowly by mutual collisions, planetary "embryos" evolve and accrete further material by impacts of planetesimals on to embryos surfaces.

The considerations above base on Safronov's planetesimals theory in empty space. On result of interest is reflected in the fact, that the Earth's core is thermally stable at the end of accretion, i.e. it is cold at the bottom and hot at the top. At the other side it is known from paleomagnetism, that the Earth's magnetic field is at least 3.5×10^9 years old, therefore soon after the end of accretion convective motions must have startet. This is inconsistent with the picture of accretion and core formation, developed above.

There are several possibilities to circumvent the difficulties, connected with a stable core. One possibility, is the existence of a primordial H—He atmosphere. A "greenhouse" effect, connected with such on atmosphere enforces high surface temperatures during the later stages of accretion. Possibly the planets are completely molten at the end of accretion and subsequent fast cooling demands convection in the whole planet.

Another possible scenario bases on late impacts of "giant planetesimals". In an ensemble of planetesimals, growing with fairly equal velocity, the "latest" planetesimals have a size of 1000 km or more. An impact of such a body onto the surface of a growing planetary embryo should melt the whole planet and push into a fast cooling state with rigorous convective motions.

Special attention deserves a third possibility a change to the other accretion scenario as it is discussed in chapter 5 of this book, namely, the "giant gaseous protoplanets". Gaseous protoplanets are in a very hot adiabatic state; fast cooling causes convection in the gas. Migration of mineral grains into the centre leads to "mineral cores", the future planets. During migration the mineral grains remain in intimate thermal contact with their gaseous surrounding. In this natural way the terrestrial planets and the rocky cores of Jupiter and Saturn establish adiabatic convective temperature distributions from the beginning. Indeed the details of thermal evolution of protoplanets and planets in this accretion scenario are unknown up to now. But this approach could be able to overcome the above mentioned difficulties.

4.3. The Giant Planets

The giant planets are the outer planets of the Solar system except Pluto, that means Jupiter, Saturn, Uranus and Neptun. Observational data of the relevant physical parameters are rather difficult to obtain. Many of them are not known with sufficient accuracy until now. But improvements in observational methods especially by space-mission led to a better understanding of their behaviour and structure. These data originate mainly from spacecraft missions for instance from the successful PIONEER and VOYAGER missions to Jupiter and Saturn. But also laboratory results dealing with the equation of state or the phase diagrams of several important elements can lead to improved models for these planets based on more complex theoretical considerations, taking into account a lot of possible processes and phenomena. For a short survey of the discovering history and a review of important data and models see for example Fink and Ip (1983). The history of some models for the interiors is reviewed shortly in Wildt (1972) or Zharkov and Trubitsyn (1980). For a more detailed view on the interiors one should be referred to Stevenson (1978) or Smoluchowski (1983).

Contrary to the terrestrial planets the mean density of the two major planets is very low; it ranges from nearly 0.7 gcm^{-3} to about 1.3 gcm^{-3}. For the other two planets it is somewhat greater: for Uranus 1.3 gcm^{-3} and Neptun 1.6 gcm^{-3}.

For Jupiter and Saturn simple calculations using the virial theorem result in central pressures in the range of about 10 TPa. In the framework of Thomas-Fermi-Dirac equation of state these planets are treated as cold bodies despite of a temperature that reaches more than 10^4 K in the centre. That results from the fact, that thermal pressure is only 1% of the potential pressure. So thermal effects may be neglected. The application of the Thomas-Fermi-Dirac theory to several elements characterized by their atomic weight and number results in mass-radius relations for such cold bodies (ZAPOLSKI and SALPETER, 1969). Such diagram is shown in Fig. 4.21. This clearly states that Jupiter and Saturn consist of a substantial amount of hydrogen. For the points are not situated exactly on the curve for H but a little nearer to that of He the planets should contain an additional fraction of helium. For model calculations commonly a composition of 75 wt. % hydrogen and 25 wt. % helium is assumed as indicated in Fig. 4.21.

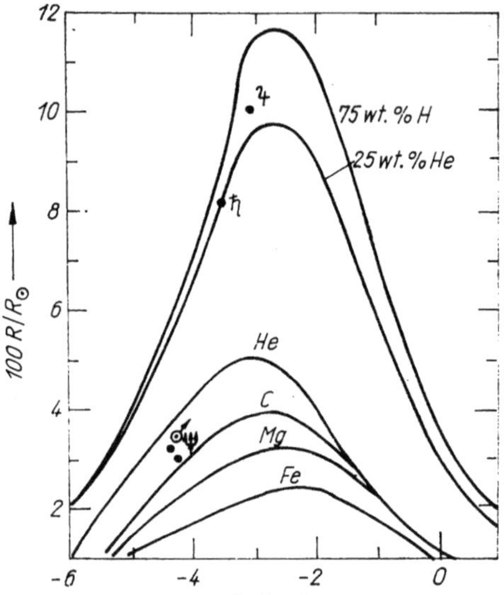

Fig. 4.21. Mass-radius relation for spherical bodies of various materials at zero temperature (adopted from ZAPOLSKI and SALPETER, 1969). (♃ — Jupiter; ♄ — Saturn; ♂ — Uranus; ♆ — Neptun)

The other two planets Uranus and Neptun have a significantly higher density than an equivalent sphere of pure hydrogen would have. It is ten times greater than for that configuration. That means, that here other elements and compounds have to be taken into account for the determination of the composition. This may be water, ammonia or methane as suggested by WILDT (1972) or COOK (1972) and ZHARKOV and TRUBITSYN (1980). Roughly spoken these two planets occupy an intermediate position between the iron and silicate containing terrestrial on one hand and the hydrogen-helium giant planets on the other. They consist to a greater extend of hydrogen compounds but also of a minor fraction of silicates and iron forming a solid core as well as pure hydrogen and helium. But it is not possible to determine the structure correctly because of the great lack of observational data and the uncertainties in the available ones. Here models based on the equation of state for relevant compounds, the theory of solids and first of all on planetogonic and cosmogonic constraints are needed.

It should be mentioned here that for Saturn and Jupiter one can expect such rocky core too.

Besides the analysis of the gravitational potential and the so deduced knowledge about the mass distribution in the interior of the giant planets the measurement of the luminosity of these planets provides important facts for the present state. The spectral power density of emission consists of a "red" and a "blue" component (HUBBARD, 1980). The blue component with a peak at a wave length of about 1 μm is reflected sunlight (effective temperature $T_e \sim 6000$ K) and may be computed from the albedo of the planetary atmosphere. Thermal radiation forms the red component. It originates from the absorption of sunradiation by the atmosphere of the planet, that heats up due to this process and reemits the radiation as thermal one. In the atmosphere an equilibrium state is established.

For Jupiter exact measurements of the infrared emission spectrum yield that the planet emits twice the energy that it receives from the Sun. That would require an inner energy source of about $5 \cdot 10^{17}$ W (STEVENSON, 1978). Saturn behaves qualitatively quite similar. It radiates more than twice the energy it absorbs.

This excess luminosity will be compared by the help of dimensionless quantities. The relative excess luminosity Q is defined with excess luminosity L, planetary age τ and gravitational energy E_g by

$$Q = \frac{L\tau}{E_g}. \tag{4.32}$$

For Jupiter a calculation results in $Q = 0.015 \pm 0.005$ and for Saturn $Q = 0.04 \pm 0.02$. The same effect is observed for Neptun. That means that all of these planets must have internal heat sources as for instance radioactive elements in the rocky core or other physical processes have to contribute to the energy balance, for example energy release due to gravitational processes as discussed by STEVENSON (1978). Another possible explanation is given in section 4.3.2.

Uranus does not exhibit such behaviour. The measured and evaluated temperatures are nearly the same what leads to the conclusion that this planet is in radiative equilibrium with the Sun.

The planets of this group possess a magnetic field as inferred from spacecraft missions and terrestrial observations of the radio emission. Data and models may be found from RUSSEL (1980). Jupiter as largest planet and that with the most rapid rotation period has the largest magnetic moment compared to all other planets. In contradiction, the field of Saturn is much smaller even in relative quantities but as a pecularity the angle between rotational axis and the dipole axis is nearly zero. Higher moments like quadrupole and octopole are very small. Meanwhile, also the existence of the magnetic field of Uranus has been verified.

Without in-situ measurements of the magnetic field of Neptun with planetary spacecrafts any assertions about this topic are rather doubtful.

The sources of these fields are expected to be dynamo actions in the interior, convective movements of electrically conductive liquid in the regions of fluid metallic hydrogen. The driving forces may be heat, precession or a solidifying core (cf. for instance BUSSE, 1975 and 1976; HIDE, 1982 and others).

SMOLUCHOWSKI (1979) pointed out that a thermally driven magnetohydrodynamic dynamo should be favoured for Jupiter and Saturn. But Jupiter's field also could be of primordial origin while this cannot be the case for the other planets of this group. A sun-driven precessional dynamo could work on Jupiter and Saturn but if present, it would be weak. The field of Uranus is connected with a thermally driven dynamo in a highly conductive hydrogen shell, while this shell is probably too thin for dynamo requirements on Neptun, so that there the field should be absent.

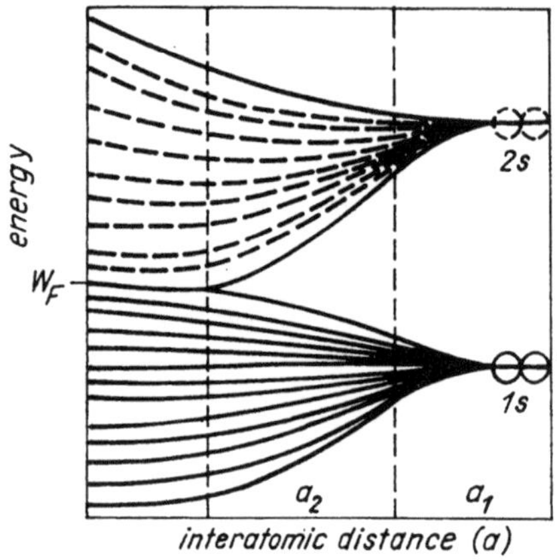

Fig. 4.22. Energy bands for He at different interatomic distances (from BLOCH, 1978). W_F — Fermi energy

Other conductive media may result from metallization processes under high pressure. It is known that for many chemical elements such pressure induced electronic transitions may occur. If for instance a crystal lattice is strongly compressed the insulated energy levels split and form bands. Figure 4.22 illustrates this for helium. At the distance a_1 He is an insulator. Increasing pressure results in smaller interatomic distances a_2. The bands overlap and He is metallized. Hydrogen should be metallic for every lattice parameter but the metallization of this element is discussed in detail in the following section.

Simple estimations for the transition pressure base on the assumption that an insulator is metallized if

$$pV \cong I \tag{4.33}$$

where I is the energy difference between the metallic and nonmetallic states and V is the atomic volume (STEVENSON, 1980).

So the dynamo of Uranus and Neptun could work with other electrically conductive media. For these two planets the conductivity of icy matter (H_2O, CH_4 and NH_3) and of compounds as SiO_2, MgO, FeS and FeO is of great importance: TORBETT and SMOLUCHOWSKI (1980) estimated, that MgO and SiO_2 have to be metallic in the whole planetary core. For FeS a transition to the metallic phase occurs already in the range of several GPa (STILLER et al., 1974). H_2O and ammonia are nearly fully ionized under these conditions and it is expected that CH_4 pyrolyses and exists in metallic phase (or probably as diamond — see ROSS, 1981).

The formation of the giant planets has some pecularities compared to that of the terrestrial planets. A scenario within the "accretional approach" has been described by SAFRONOV (1969 and 1972), who shared this process into several stages that were characterized by different accretional mechanisms and gas-dust interactions, but several difficulties arise from these assumptions. CAMERON (1978) proposed another model — the growth and evolution of giant protoplanets. The present model which is on the line of a protoplanetary origin is worked out in part 5.

The outermost double-planet of the Solar System, Pluto/Charon should be described only briefly. This planet is quite different from the giant planets. The observational base is exceedingly small and therefore it is very difficult to obtain reliable data.

The earlier proposition of a terrestrial similarity has been rejected according to new observations. The density is now assumed to lie in the range below about 1.5 gcm^{-3} (FINK and IP, 1983) but this is not well constraint. That excludes a rocky planet. From the observed spectrum the presence of methane is deduced but it is not quite clear if it is gaseous or solid. But relevant parameters, cosmochemical and planetogonic aspects imply, that Pluto consists mainly of icy matter, for instance a mixture of methane, water and ammonia because the density produced by slightly compressed pure CH_4 ($\varrho \sim 0.5$ gcm^{-3}) is insufficient to explain the observed density. There are no data for the magnetic field available and the inner structure is unknown. So a lot of experimental and observational work has to be done to improve the knowledge about this planet significantly (but for the giant planets as well).

The satellites and moons of the outer planets and their particular structures and properties should not be discussed here. Therefore one should be referred to the special literature (e.g. STEVENSON, 1978).

4.3.1. Phase Diagram of Hydrogen

As already mentioned the most abundant element in the planetary interiors of Jupiter and Saturn is hydrogen with a small admixture of helium (for model calculations usually 75 wt. % H and 25 wt. % He). To determine the thermal state of the interior it is necessary to be familiar with the properties of the elements under appropriate conditions of temperature and pressure.

Here the phase diagram of pure hydrogen is discussed, based on considerations and calculations of ZHARKOV and TRUBITSYN, HUBBARD and SMOLUCHOWSKI.

To include all relevant regions from the atmosphere to the deep interior of these two planets, this diagram must comprehend the whole region from ideal gas behaviour up to dense degenerate matter.

Figure 4.23 shows such a schematic phase diagram (cf. for instance STEVENSON, 1978; ZHARKOV and TRUBITSYN, 1980 or SMOLUCHOWSKI, 1983). Some aspects still remain uncertain for example the position of the critical point for the dielectric-metal transition (STEVENSON and SALPETER, 1977; FRANCK, 1980, 1981a and 1981b; EBELING and RICHERT, 1985) or the existence of the metallic solid phase.

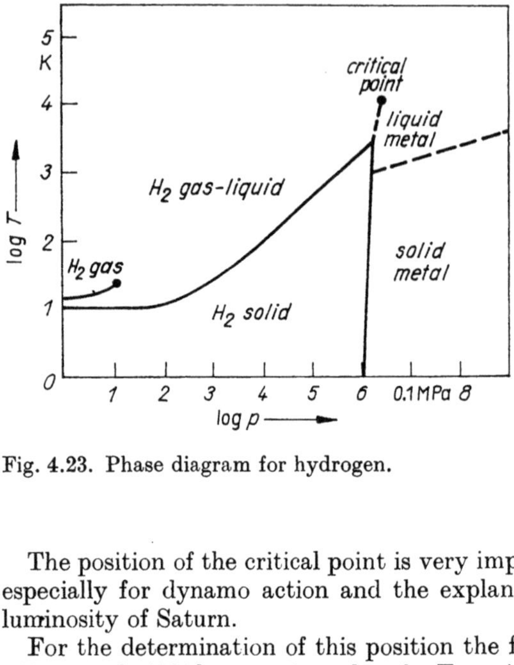

Fig. 4.23. Phase diagram for hydrogen.

The position of the critical point is very important for the planetary models especially for dynamo action and the explanation of the anomaly in excess luminosity of Saturn.

For the determination of this position the following model (FRANCK, 1980; 1981a and 1981b) was introduced: Experimental results and theoretical estimations show, that the metallization is expected to occur at densities of about 1 gcm^{-3}, the transition pressure lies in the region 200...300 GPa (HAWKE et al., 1978). Under these conditions the intermolecular distance becomes comparable to the size of the molecules. That results in strong intra- and intermolecular interactions.

Figure 4.24 illustrates the model. The system of highly compressed hydrogen is divided into separate cells with only one proton in every cell. The position of that proton corresponds to a lattice position. This proton at position i may be surrounded by an electron that is only bound to that proton (atomic dielectric hydrogen) or simultaneously to another neighbour proton (molecular dielectric hydrogen). These states are called dielectric (d). If there is no single electron bound to that considered proton the state is called metallic (m).

This model leads to the quantumstatistical problem of a particle with two

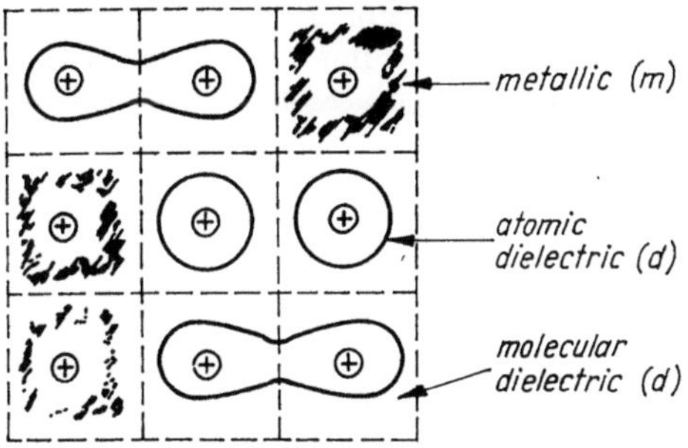

Fig. 4.24. Model for the metallization of hydrogen.

possible states. The describing Hamiltonian can be constructed from creation and annihilation operators a^{i+} and a^i, respectively in the framework of the second quantization

$$H = \sum_{\alpha} E_\alpha a_\alpha{}^{i+} a_\alpha{}^i + \sum_{\alpha\beta\gamma\delta} v^{ij}_{\alpha\beta\gamma\delta} a_\alpha{}^{i+} a_\beta{}^i a_\gamma{}^{j+} a_\delta{}^j \tag{4.34}$$

where E_α — single particle energy; $v^{ij}_{\alpha\beta\gamma\delta}$ — interaction potential; α, β, γ, δ are the two possible states: metallic or dielectric.

With this Hamilton-operator it is possible to define some spin operators (for special calculations one should be referred to the literature cited above). The operator S_i^z characterizes the state of cell i: $S_i^z = \frac{1}{2}$ means that the cell is in the metallic state while $S_i^z = -\frac{1}{2}$ dielectric. The mean value of this operator $\langle S^z \rangle$ changes according to the difference between the energies of the metallic and dielectric phases (Δ). At low temperatures this change takes place abruptly while at high temperatures it is continuous (cf. Fig. 4.25).

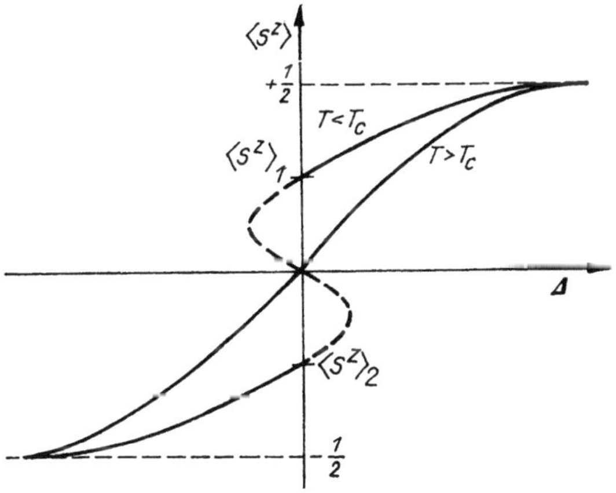

Fig. 4.25. Dependence of the mean value $\langle S^z \rangle$ on the energy difference Δ between metallic and dielectric phases.

At a certain temperature T_c this jump vanishes. For $T > T_c$ there is no abrupt change of properties; a phase boundary does not exist. The critical temperature may be evaluated from the equation

$$k_B T_c = \frac{hf}{\ln\left[\frac{1 + 2(hf/J_0)}{1 - 2(hf/J_0)}\right]}, \tag{4.35}$$

with h — Planck's constant; f — frequency for the interchange between dielectric and metallic sites and J_0 — quantity characterizing the interaction between the two phases.

The interchange frequency may be estimated from simple considerations on the diffusion in liquids (FRENKEL, 1957b). It is approximately about $1.3 \cdot 10^{14}\ s^{-1}$ in the case of metallic and dielectric hydrogen. Numerical values for J_0 may be found from papers by TRUBITSYN (1966) and ROSS (1974). If both phases have nearly the same energy $\left(\sum_j v^{ij}_{mmmm} = \sum_j v^{ij}_{dddd}\right)$ approximative values may be taken from NEECE et al. (1971) and KAGAN et al. (1978); the interaction energy of a dielectric particle in metallic hydrogen $\left(\sum_j v^{ij}_{dmmm}\right)$ is estimated from calculations of the energy of a hydrogen atom or molecule in jellium by NORSKOV (1979).

All different models lead to the same conclusion that the critical point is situated in the region about $6 \ldots 9 \cdot 10^3$ K for the phase boundary between liquid metallic and dielectric hydrogen. Other authors (for instance EBELING and RICHERT, 1985 and others cited there) reach with other methods at significantly higher temperatures.

The consequences of the existence of a critical point in the region $6 \ldots 9 \cdot 10^3$ K will be discussed in the next section.

4.3.2. Models for the Planetary Interior

The models for the interior of the giant planets were successfully improved in the last years because of an increase in available data from spacecraft missions for instance the PIONEER and especially the VOYAGER projects. Several years ago the question if the giant planets were in a solid or liquid state remained unsolved. Many efforts have been made by V. N. ZHARKOV, V. P. TRUBITSYN, W. B. HUBBARD und R. SMOLUCHOWSKI for the development of the gas-liquid adiabatic model.

STEVENSON (1978) describes the input parameters needed for the model construction:

(i) the planetary rotation period,
(ii) the equation of hydrostatic equilibrium,
(iii) the equation of state $P = P(\varrho, T, X_i)$, with $X_i = X_i(r)$ — fractional abundance of constituent i, and the phase diagrams of the components,
(iv) internal heat sources with appropriate distribution,

(v) thermal transport properties (conduction, convection, radiation); these determine the pressure-temperature relationship,
(vi) theory of planetary figures.

The models have to fit the observational data as mass, radius, gravitational moments as well as observed composition and thermal state and the evolutionary and cosmochemical constraints. They should result in predictions of the internal structure, the pressure-temperature relationship with possible phase transitions and of generating mechanisms of the magnetic field.

At first the interiors of the two major planets Jupiter and Saturn shall be discussed. As already has been pointed out these consist of a small rocky core and a hydrogen-helium mixture in metallic and non-metallic states. The core is composed of silicates and ferrous materials and possibly ice that forms a separate layer around the core in some models. Its density reaches nearly 5.5 gcm^{-3} according to MIZUNO (1980), in SLATTERY's model 22 gcm^{-3} and 8.5 gcm^{-3} for Saturn from the ANDERSON et al. (1980) model. The size is about 15...20% of the planetary radius, the volume fraction is less than 1%. MIZUNO (1980) showed that the core mass of all giant planets should be about 10 Earth masses. Recent models predict a slightly higher mass. For instance SLATTERY (1977) proposes for Jupiter $14 M_{\oplus}$ while ANDERSON et al. (1980) get $16.5 M_{\oplus}$ for Saturn. But it is evident that this mass depends strongly on

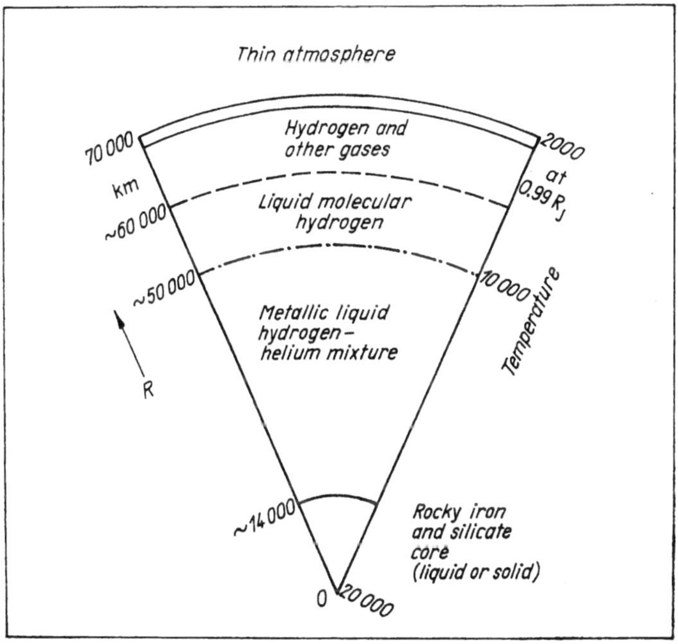

Fig. 4.26. Internal structure of Jupiter (adopted from STEVENSON, 1978 and COLE, 1981).

the model. This rocky fraction may be in a solid or liquid state and probably should be metallized by the high central pressures expected there. The core is assumed to be isothermal.
The H-He envelope is structured by different layers as molecular and metallic hydrogen, helium and possibly ice.

Figures 4.26 and 4.27 show the interior of Jupiter and Saturn, respectively.

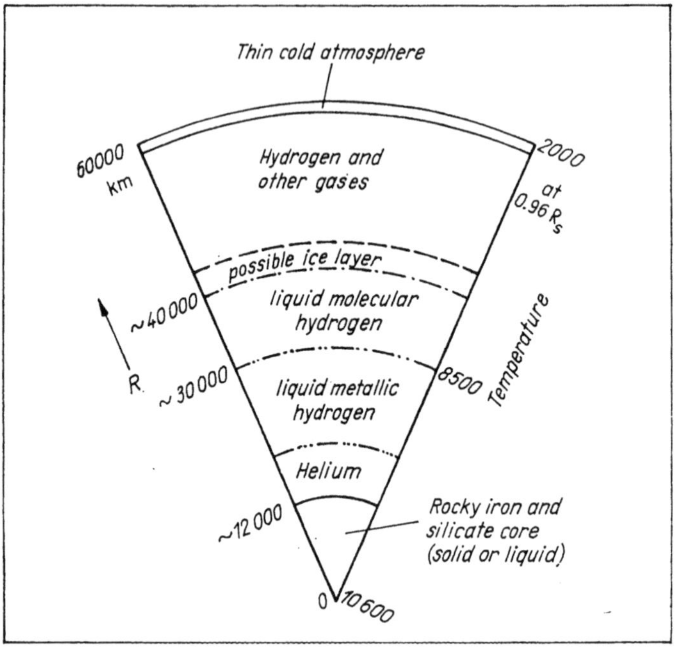

Fig. 4.27. Internal structure of Saturn (adopted from Stevenson, 1978 and Cole, 1981).

These models shall now be regarded in connection with the phase diagram of hydrogen. The critical point for the metal—non-metal transition of liquid hydrogen is applied to both of the planets Jupiter and Saturn. In Fig. 4.28 this diagram is completed with typical Jovian and Saturnian adiabats. It can be seen that for Jupiter there is no phase boundary between liquid metallic and dielectric hydrogen because the adiabat lies above the critical point. That agrees with a hypothesis by Smoluchowski (1975), that large-scale convective dynamo action extends from the regions of metallic hydrogen to outer layers with the dielectric modification. The great amount of highly electrically conductive fluid that permits operation of a hydrodynamic dynamo is therefore much closer to the surface. This accounts for the great strength of the observed higher-than-dipolar moments of the magnetic field (Smoluchowski, 1983).

In Saturn this dynamo action is presumably restricted to the inner regions with liquid metallic hydrogen. This planet possesses some further pecularities.

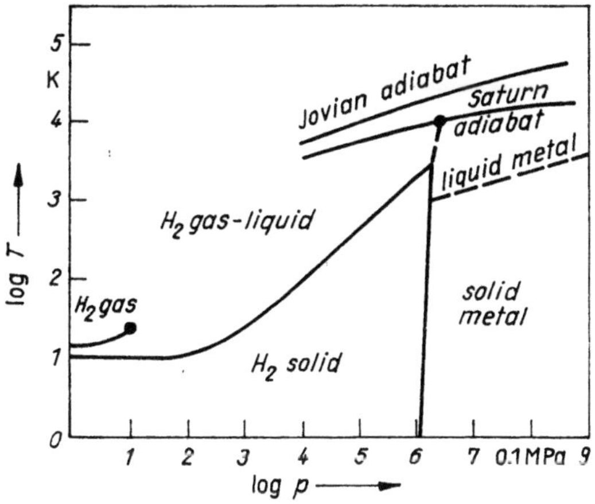

Fig. 4.28. Phase diagram of hydrogen completed with planetary adiabats.

So the planet's luminosity is higher than it could be considering its primordial heat and the known rate of cooling.

Two possible energy sources should be discussed here. The first one can be related to the phase transition. In the framework of homogeneous thermal evolution (STEVENSON, 1978) the excess luminosity L is estimated from the secular variation of the internal thermal energy by cooling

$$L = -\frac{\mathrm{d}}{\mathrm{d}t}\left[\frac{4\pi}{3} R^3 c_v T_i\right]. \tag{4.36}$$

R is the planetary radius, c_v the specific heat per unit volume and T_i a characteristic inner temperature. Neglecting special effects one can assume that only variations of T_i with time are important in (4.36). So this equation becomes approximately

$$L \simeq -\frac{4\pi}{3} R^3 c_v \frac{\mathrm{d}}{\mathrm{d}t} T_i, \tag{4.37}$$

what is correct in that form only for Jupiter. For Saturn the phase transition has to be taken into account, especially the behaviour of the specific heat. The mean field theory predicts a peak of $3/2k_B$ (k_B — Boltzmann's constant) for c_v at the critical temperature T_c. More recently STANLEY (1973) proposes divergencies of the form

$$c_v \sim (T_c - T)^{-\alpha'} \tag{4.38}$$

with the critical index α' ranging from 0.1 to 0.6. Figure 4.29 describes this behaviour. At the critical point the specific heat increases significantly and is strongly temperature dependent compared to the normal case. FRANCK (1981a)

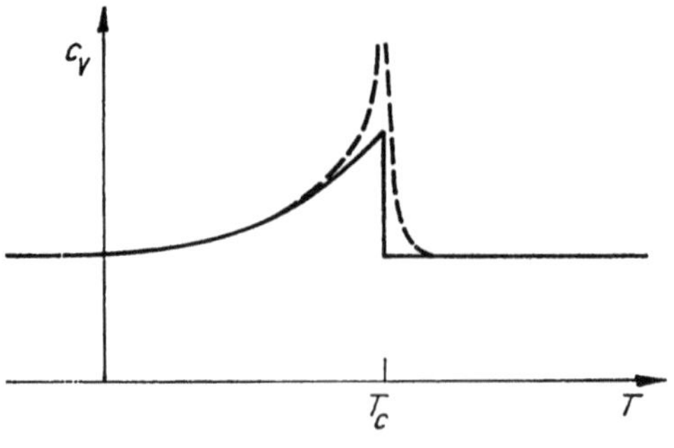

Fig. 4.29. Specific heat near a phase transition (broken line: recent theory).

calculates an additional contribution ΔL to the excess luminosity that results from a layer ΔV with $T_i \lesssim T_c$

$$\Delta L \lesssim \Delta V \frac{\mathrm{d}T_i}{\mathrm{d}t} T_i \frac{\mathrm{d}c_v}{\mathrm{d}T_i}. \tag{4.39}$$

This contribution may be the reason for the anomalous high excess luninisity of Saturn compared to Jupiter. Saturn cools faster and releases more energy at the same temperature decrease than Jupiter.

The second reason for the excess luminosity is connected with the phase diagram of a hydrogen-helium mixture. Experimental results by STRETT (1974) and investigations by STEVENSON and SALPETER (1976) (see also HUBBARD, 1972) indicate that helium has a much smaller solubility in dielectric and also metallic hydrogen under pressures of several 100 GPa. So in a mixture of 90% hydrogen and 10% helium (referring to the particle densities) at temperatures below 10000 K a phase separation into a helium-rich and a helium-poor phase will occur (cf. Fig. 4.30). Due to this separation helium-rich droplets will form and after reaching a radius of approximately 1 cm these sink into the

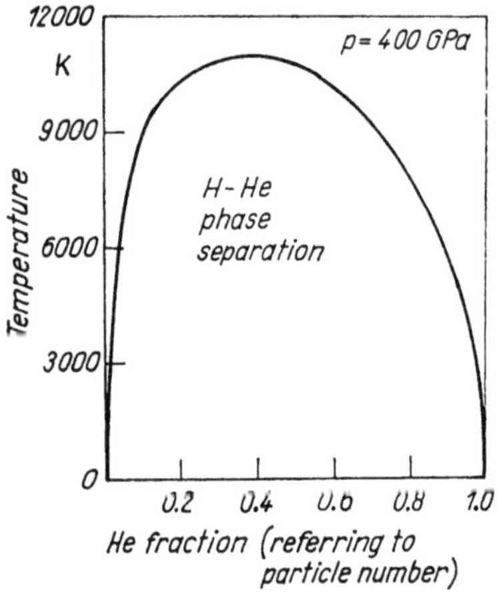

Fig. 4.30. Phase separation in a H—He mixture (from STEVENSON and SALPETER, 1976).

planetary interior under the influence of the gravitational field. The energy is released by viscous dissipation. In deeper regions according to temperatures above the separation temperature these droplets dissolve. Convective interactions deplete outer layers and the atmosphere of helium that is enriched in the interior.

This process can take place in Saturn because of the lower temperature and pressure. These are lower than the critical data for the miscibility gap. So the planetary interior from Fig. 4.27 has to be completed by Fig. 4.31. As Smoluchowski (1983) emphasizes this process of helium rain is self-regulating. He estimates the energy release for those droplets drifting across the 500 km layer to be about $2 \cdot 10^8$ Jkg^{-1}.

The results of the VOYAGER mission support the helium rain model because in the atmosphere there was measured a lower He content than for Jupiter. So this model should be preferred.

For other relevant data as for example pressure-temperature diagrams in relation to the radius, thermodynamic data and the mathematical formalism for constructing the models see Stevenson (1978), Smoluchowski (1983), Zharkov and Trubitsyn (1980).

Models for the other two giant planets Uranus and Neptun are not well constraint because of the lack of available data. Until now there exist no spacecraft data from direct observations of these planets. The main constitu-

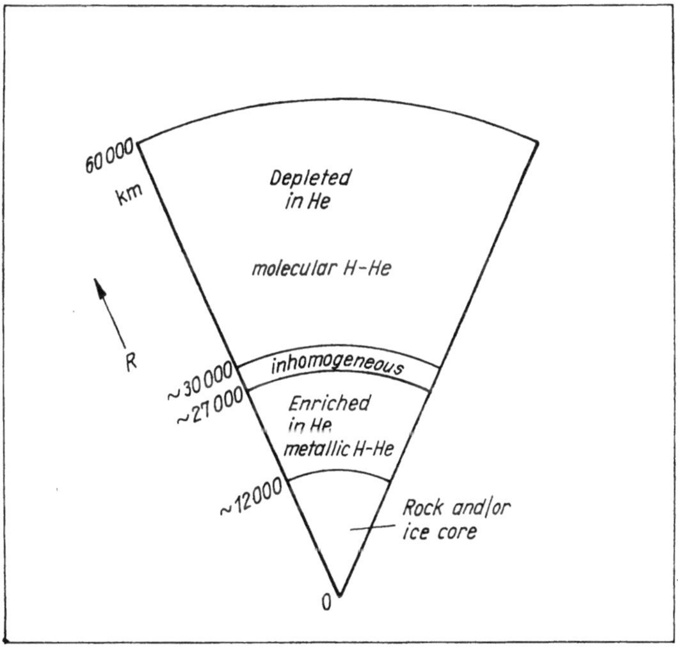

Fig. 4.31. Differentiation in Saturn's interior (from Stevenson, 1982a).

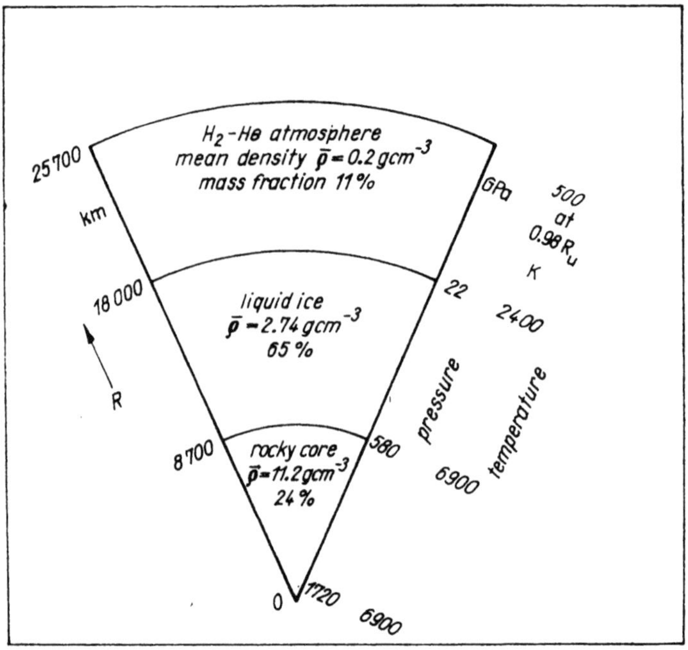

Fig. 4.32. Model for Uranus (from FINK and IP, 1983).

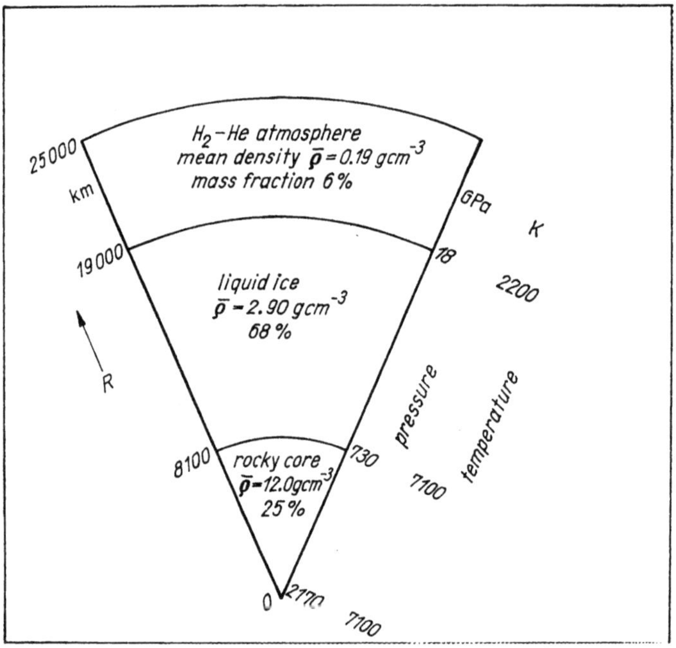

Fig. 4.33. Model for Neptun (from FINK and IP, 1983).

ents are rock (SiO_2, MgO, FeS and FeO), ices (H_2O, NH_3, CH_4) and a H—He mixture in the atmosphere. Very rough models are depicted in Figures 4.32 and 4.33.

Several other models as for instance the enrichment of Uranus' atmosphere in methane (Hubbard and Mac Farlane, 1980) with the appropriate pressure-temperature diagrams are discussed by Stevenson (1978) as well as Zharkov and Trubitsyn (1980) or Smoluchowski (1983).

For Pluto, the outermost planet, there exist no reliable models because of a great deficiency of observational data.

Planetogonic Problems

The present state of the interior of the giant planets indicates that these planets were all much hotter when they formed and that they all have rock and/or ice cores of ten to thirty Earth masses:

At this time two formation scenarios are discussed (Stevenson, 1982b): one based on the purely gaseous instability hypothesis (such as Jeans collapse) and one on the nucleated instability hypothesis, in which a core of rock and/or ice initiates a hydrodynamic collapse of the surrounding gaseous envelope. The second hypothesis is favoured by Stevenson because it can explain the remarkable similarity of the core masses despite the wide range of total planetary masses. In this way condensation, sedimentation and accretion may have begun much earlier in the outer parts of the solar nebula, where the temperatures were lower.

In our opinion, the prior formation of the outer planets again indicates that quicker scenarios should be favoured because only in this case the temporal difference in the formation of terrestrial and giant planets can be established without problems.

5. Planetogonically Relevant Physical Processes and Parameters

D. Möhlmann

It is the aim of this part to discuss planetogonically relevant physical processes on the basis as well of astronomical observations as of the mathematical description of these physical processes. As will be shown in chapter 5.1., there is an evolution of interstellar matter towards growing protostars with surrounding preprotoplanetary matter. This preplanetary matter may accumulate in a relative thin disk due to the possible action of large-scale protostellar magnetic fields and by mechanical processes (see chapter 5.2.). This disk consists mainly of plasma, gas, and grains and growing small solid bodies. A part of the matter of the disk may evolve also later towards small solid bodies. Such disks can become gravitationally unstable in the large scales (in early phases) and in the small scales (at later phases of lower density), evolving through intermediate bands or rings towards protoplanets as the predecessors of planetary sized bodies and directly towards smaller solid bodies in a later phase. Physical processes, relevant for this planetogonic scenario are described within this part.

5.1. Pre-main Sequence Evolution of Circumprotostellar Matter

In order to get a better understanding of possible initial and boundary conditions for the description of evolution of planetary bodies from preplanetary matter, what may evolve from circumprotostellar matter, it is necessary to derive a rough estimation of the physical processes and relevant parameters, connected with the evolution of young stars and their environment from interstellar matter. These processes of star formation may quite naturally be the cause for circumprotostellar matter to evolve as the predecessor of preplanetary matter and planets. Therefore it is aim of this chapter to characterize qualitatively the evolution towards young stars and their environment. This is based mainly on discussing the physics of these evolutionary processes and on reviewing the literature on star formation, but with the restriction to relatively low mass stars with masses less than a few solar masses. There are two causes for this restriction, namely at first, to have the possibility to compare the resulting planetogonic conclusions with the properties of our planetary system, and secondly, the evolution of more massive protostars exceeding a few solar masses, differs essentially with respect to that of protostars with lower masses. This takes place due to the fact, that if a collapsing protostellar core exceeds approximately $(3-5)\,M_\odot$, it may contract further and begins nuclear burning while continuing to accrete matter. It is then the internal luminosity due to nuclear reactions in the core which determines the structure of the protostar and its environment, and not, as in the lower mass case, the luminosity due to accretion of the protostar. Consequently, also the circumstellar environments, which are of planetogonic interest, evolve on different ways in these two cases.

Here, the word "protostar" needs some definition. Throughout this book a protostar is a gravitationally bound and collapsing or contracting mass accumulation (fragment) which does not fragmentate during its further evolution towards a mainsequence star. The energy source of protostars is gravitational energy due to the collapse.

For the sake of a precise wording, characterizing the regions of matter, surrounding stars (circumstellar matter) or protostars (circumprotostellar matter) the following terminology shall be used throughout this book:

a) environment or envelope — including all surrounding matter (infalling or orbiting as well as leaving the gravitational sphere of influence),
b) shell — the hot (plasma-)region within a few stellar (or protostellar) radii,
c) disk — the outer and relatively cooler and more dusty region out of the shell, which is concentrated close to a plane.

This chapter on pre-main sequence evolution of protostars and circumprotostellar matter states that there is today a qualitative agreement between predictions from calculations and observed properties. Therefore, and in order to have a basis for the following chapter on the evolution of circumprotostellar matter towards protoplanets, the essential physical properties of environments which are compatible with theory and observation are summarized at the end of the chapter.

5.1.1. Temperature and Density of Prestellar Matter

It is generally accepted today that stars evolve from cold dark dense interstellar clouds of gas and dust with mass densities in the range from 10^{-18} kg · m^{-3} to 10^{-16} kg · m^{-3} and total masses between 10^2—10^4 solar masses. These clouds are opaque to visual and ultraviolet radiation while infrared radiation is yet able to escape at these densities. Consequently, the energy balance is different from that usually considered for the optically thin interstellar medium. A detailed consideration of the heating and cooling processes would be out of the scope of this book, but in order to have some links as well to chemical and mineralogic evolution in prestellar matter, what has been discussed in chapter 3, as to the initial conditions for the further collapse towards protostars, some properties of the prestellar matter shall be summarized.

The main heating processes are cosmic ray heating (Goldsmith et al., 1969) and compressional heating if the selfgravitation is sufficiently high for a gravitational collapse. The dominating cooling processes are collisional excitation of CII ions and O I ions by H atoms, followed by the emission of infrared radiation and inelastic collisions between gas molecules and dust grains and further followed by infrared emissions from the dust grains (Hayashi, 1966; Goldsmith, 1969). Following Larson (1973), Table 5.1. gives for a

Table 5.1. Equilibrium temperature in dense interstellar clouds (from LARSON, 1973)

	$M = M_0$	$M = 10^3 M_0$
$\log \varrho$ (kgm^{-3})	T (K)	T (K)
−19	50.6	48.1
−18	25.4	24.6
−17	15.6	16.3
−16	8.4	6.6
−15	5.1	5.1
−14	5.4	5.4
−13	6.3	6.3
−12	7.5	7.5
−11	9.1	9.1
−10	11.0	11.0
−9	13.3	13.3

collapsing cloud of 10^0 and 10^3 solar masses the values for cloud density and temperature. Here, the above mentioned energy balance has been taken into account explicitely.

As will be discussed later, the evolving protostar becomes opaque also to infrared radiation at densities around 10^{-10} kg · m^{-3}. By the way, these densities are characteristical too for the outer preplanetary disk, which therefore was able to cool out and to remain at relatively low temperature (see chapter 5.3.). For protostellar evolution the resulting increase of temperature within the now opaque region indicates a qualitatively new stage of the collapse following the isothermal compression, as it is described by Table 5.1. The resulting consequences for protostellar evolution will be discussed in more detail in section 5.1.5.

The dominant heating and cooling processes at low densities (below $3 \cdot 10^{-17}$ kg m^{-3} for $M = M_\odot$ and $3 \cdot 10^{-18}$ kg m^{-3} for $M = 10^3 M_\odot$) are cosmic ray heating and C II cooling. At higher densities compressional heating and grain cooling become dominant (LARSON, 1973; HAYASHI, 1966). The free escape of infrared radiation is the cause for the resulting isothermal character of the collapse in these early stages.

Temperatures, well comparable to the derived values are observed in many of the dark clouds where star formation is believed to take place (HEILES, 1971; EVANS, 1981). The densities within these clouds are usually of the order of 10^{-17} kg m^{-3}, or even higher, and the temperatures range from about 5 K to 20 K. These observational results strengthen the above given conclusion that the early collapse of a protostar and also the outer regions in later "planetogonic" phases are approximately isothermal with temperatures of about 10 K. The inner (opaque) preplanetary regions should be of a different state. This is mainly due to the heating of these inner regions.

5.1.2. The Jeans Criterion

Stability of interstellar clouds against gravitational collapse is caused mainly by the (possibly coupled) action of internal and external thermal pressure, centrifugal forces and magnetic fields. As a rough mathematical formulation for a stability criterion the virial theorem can be used, which states that for an isolated (spherical) configuration to be stable the collapse supporting energy E_G of selfgravitation and of a possible external pressure have to balance twice the internal (kinetic) energies:

$$\frac{3}{5}\gamma\frac{M^2}{R} + 4p_e\pi R^3 \approx \frac{4}{5}M\omega^2R^2 + 3\frac{MkT}{\mu m_H} + \frac{4}{3}\pi R^3\frac{B^2}{\mu_0}. \tag{5.1}$$

The gravitational energy of a spherical cloud of mass M, volume $V = \frac{4}{3}\pi R^3$ and radius R can be found from

$$E_G = \gamma\int_0^M \frac{m}{r}\,dr = \frac{\gamma}{3}(4\pi\varrho)^2\int_0^R r^4\,dr = \frac{3}{5}\gamma\frac{M^2}{R}, \tag{5.2}$$

where $\gamma = 6.674 \cdot 10^{-11}\,kg^{-1} \cdot m^3 \cdot s^{-2}$ is the constant of gravitation. The kinetic energy E_R of rotation of a spherical cloud, assumed for simplicity to rotate with a uniform angular velocity ω, is given by

$$E_R = \frac{2}{5}M\omega^2R^2. \tag{5.3}$$

For the internal thermal energy E_{th} of a cloud of temperature T and mean molecular weight μ the representation

$$E_{th} = \frac{3}{2}\frac{MkT}{\mu m_H} \tag{5.4}$$

is sufficient for not to dense gaseous matter. Here, $k = 1.380 \cdot 10^{-23}$ Ws deg^{-1} is Boltzmann's constant, $m_H = 1.672 \cdot 10^{-27}$ kg is the mass of the hydrogen atom.

The last term in equ. (5.1) describes the internal magnetic energy E_M, possibly involved in the cloud. The influence of internal magnetic fields on the dynamics of interstellar clouds is still a matter of debate. This will be discussed in more detail in section 5.1.4.

A quantitative verification of equ. (5.1) can be found from astronomical observations of interstellar clouds (v. Hoerner, 1975), giving values for energy densities in the range

$$\bar{\eta} \approx (10^{-13} - 10^{-14})\,\frac{Ws}{m^3} \quad \text{for all the above} \tag{5.5}$$

mentioned energies.

In order to get a first estimation of the physics involved in the virial theorem, the circumstances under which selfgravitation will dominate over the expanding action of internal thermal pressure and cause collapse to occur, will be determined with the assumption that all other forces can be neglected as a first approximation.

This gives a formulation of the Jeans criterion

$$E_G > E_{th}, \quad \gamma \frac{M}{R} \geqq 5 \frac{kT}{\mu m_H}, \tag{5.6}$$

where the precise extend to which E_G has to exceed E_{th} can not be determined generally (as a universal constant). This is due to its dependence on model assumptions for the thermal evolution of the collapse. The range of the numerical constant in the JEANS criterion is given at one side by the virial theorem (upper limit) and on the other side by numerical model calculations (LARSON, 1969), giving a lower value of 2.17 instead of 5. As there is no essential physics in the precise determination of this numerical value, this problem shall not be discussed further. The value 2.5 will be used throughout the following. This value corresponds to an equilibrium between gravitational and internal thermal energy.

The representation (5.6) of the JEANS criterion can be reformulated to give the critical radius from which stable equilibrium is no longer possible, yielding

$$R_J = 0{,}4 \frac{\gamma M}{kT/\mu m_H}, \tag{5.7}$$

or with $M = \frac{4}{3} \pi R^3 \varrho$

$$R_J = \left(\frac{3k}{1{,}6\pi\gamma\mu m_H} \frac{T}{\varrho}\right)^{1/2}. \tag{5.8}$$

The JEANS criterion can be written also to give the minimum mass M_J which a cloud of given density and temperature must have in order to collapse gravitationally

$$M_J = \left(\frac{5}{2} \frac{kT}{\gamma\mu m_H}\right)^{3/2} \cdot \left(\frac{3}{4\pi\varrho}\right)^{1/2} = 1.93 \left(\frac{k}{\gamma\mu m_H}\right)^{3/2} \cdot \varrho^{-1/2} \cdot T^{3/2}. \tag{5.9}$$

Table 5.2 gives the numerical values for M_J and R_J for the collapsing cloud (of hydrogen) of $1 M_\odot$, as it has been described by Table 5.1.

Two interesting results can be derived from the estimations given with Table 5.2:

— At the end of the isothermal collapse phase the Jeans mass approaches the range of planetary masses, indicating that under appropriate conditions planetary sized bodies could evolve. As will be discussed in section 5.1.5., the central mass is relatively small at this time. Consequently, tidal actions were not important, they did not present any constraints to the formation of these "first generation" planetary bodies (brown dwarfs?). These bodies have to take part in the inward directed motions due to the collapse.

— At the beginning of a gravitational collapse of a dense interstellar cloud, the minimum unstable mass predicted by the JEANS criterion is of the orders of 10^3 solar masses of more. This indicates that single stars can not have been formed directly by a direct collapse of an interstellar cloud. In order to get stars of normal mass, densities of several orders of magnitude higher than those observed in the interstellar medium are necessary. Consequently, star formation has to take place from fragments with preenhanced densities, which could be formed in the course of the collapse of the more massive interstellar clouds. The detailed physics of this fragmentation is not yet well understood, but there is no doubt that the collapse of dense interstellar clouds is quite naturally be connected with local mass-accumulations which are gravitationally bound and which begin to collapse by themselves. The reason for this is that the free fall time of a collapse varies as $\varrho^{-1/2}$ (see section 5.1.5.). Consequently, denser regions always collapse faster and any small density fluctuation is rapidly amplified, provided that their mass exceeds the JEANS mass M_J, as it is given by equ. (5.9). Following an idea of BODENHEIMER (1978), it is in actual discussion that several (cascadic) stages of fragmentation may occur which could additionally lead at least also to the formation of protostars with masses of the order of the solar mass $M_\odot = 1.99 \cdot 10^{30}$ kg.

This scenario, which is well founded from the point of view of the involved physical processes, and which can be characterized as a collapse caused fragmentation with resulting formation of a number of protostars from the same interstellar cloud, is in good agreement with astronomical observations which indicate that young stars appear in associations or groups. The fragmentation process will be discussed in more in the next section.

On the other side, it seems also to be possible that under special conditions protostars can be formed as a result of thermal instabilities in a medium which is in a cooling process from high temperature (HUNTER, 1969; STEIN et al.,

Table 5.2. The Jeans mass and radius as functions of density and temperature (from LARSON, 1973)

$\log \varrho$ (kgm^{-3})	T (K)	$M_J(M_0)$	R_J (pc)
−19	50.6	$1.5 \cdot 10^3$	6.3
−18	25.4	$1.7 \cdot 10^2$	1.4
−17	15.6	$3.0 \cdot 10^1$	$3.5 \cdot 10^{-1}$
−16	8.4	$3.2 \cdot 10^0$	$8.1 \cdot 10^{-2}$
−15	5.1	$4.9 \cdot 10^{-1}$	$2.0 \cdot 10^{-2}$
−14	5.4	$1.7 \cdot 10^{-1}$	$6.5 \cdot 10^{-3}$
−13	6.3	$6.7 \cdot 10^{-2}$	$2.2 \cdot 10^{-0}$
−12	7.5	$2.7 \cdot 10^{-2}$	$7.6 \cdot 10^{-4}$
−11	9.1	$1.2 \cdot 10^{-2}$	$2.7 \cdot 10^{-4}$
−10	11.0	$4.9 \cdot 10^{-3}$	$9.2 \cdot 10^{-5}$
−9	13.3	$2.0 \cdot 10^{-3}$	$3.2 \cdot 10^{-5}$

1972), from thermo-chemical instabilities due to CO and H_2O formation-destruction (GLASSGOLD and LANGER, 1976) or to H_2 formation-destruction (HUTCHINS, 1976), or from the action of an external pressure. JURA (1976) has shown numerically that galactic spiral density shock waves can trigger the collapse of clouds with lower than the JEANS mass, and that especially supernova caused shock waves may cause a density enhancement, being on this way the cause for collapsing clouds with $M \sim M_\odot$. In this connection, the results of DORFI (1981) should be mentioned, indicating a modifying action of magnetic fields on the collapse, favouring the formation of highly non-homologous, bar-like structures (which may continue to collapse). A further remarkable result of these calculations is to explain, that due to the action of the magnetic field only a few percent of the collapsing cloud may evolve to protostars. This is in agreement with the results of ZUCKERMAN and PALMER (1979), that the stellar birth rate should by 30–100 times greater, if all matter of the interstellar clouds would be converted to stars.

In spite of the fact that the above mentioned initial conditions are rather different from those derived from the JEANS criterion, the early stages of the resulting protostellar collapse are qualitatively very similar in all of the discussed above approaches (STEIN et al., 1972). This is an essential justification for the limitation to study the processes of the formation of early stars and their environment within the frame of the collapse of only one fragment. The results of this single-fragment-approach will be discussed in section 5.1.5.

5.1.3. Rotational Fragmentation

The JEANS criterion has been derived in the foregoing section without taking into account the stabilizing action of rotation and angular momentum conservation. These influence shall be discussed now. The kinetic energy E_R of rotation of a uniformly rotating spheroidal cloud of uniform density has been given by equ. (5.3).

Instead of equ. (5.6) it follows now as a condition that collapse will occur

$$\frac{3}{5}\gamma\frac{M^2}{R} > \frac{2}{5}M\omega^2R^2 \quad \text{or} \quad v = \omega R < (3\gamma M/2R)^{1/2}. \tag{5.10}$$

This implies for the order of magnitude of the angular velocity

$$\omega < (2\pi\gamma\varrho)^{1/2} \tag{5.11}$$

while the characteristic radius for equilibrium can be found to be

$$R = \frac{\frac{3}{5}\gamma M^2}{E_R + E_{th}}, \tag{5.12}$$

giving the order of magnitude of the maximum extension of the possibly remaining circumprotostellar environment, if it has not been compressed by external pressure.

The ratio between rotational and gravitational energy can be shown to be

$$\frac{E_{\mathrm{R}}}{E_{\mathrm{th}}} = \frac{25L^2}{3\gamma M^3 R} \tag{5.13}$$

where $L = \frac{1}{5} M\omega R^2$ has been used for the total angular momentum (spin) of the cloud. The angular momentum remains unchanged (conserved) throughout the collapse.

The main conclusions, which also result from numerical axisymmetric (two-dimensional) models are that a cloud must be more compressed (smaller R-values in equ. (5.12)) if it is rotating ($E_{\mathrm{R}} > 0$) than a nonrotating cloud if gravity has to dominate over the combined action of pressure and centrifugal force, and that the ratio of centrifugal to gravitational forces increases as the collapse proceeds (see equ. (5.13)). Since the collapse proceeds faster in the denser central regions, the centrifugal force in that regions becomes first large enough to balance the action of gravity and to decelerate and concentrate further collapse toward the axis of rotation. There a ring should form which increases in mass due to the inflow of matter from outside where centrifugal force has not yet become large enough to balance gravity (Fig. 5.1a).

A doughnut-like configuration with a central density minimum should evolve (LARSON, 1973), as it has been illustrated by Fig. 5.1b.

It shall be noted here that independent of rotation, collapse is still possible along the axis of rotation yielding a further increase in central mass.

It has been suggested and modeled numerically (OSTRIKER and BODENHEIMER, 1973; BODENHEIMER, 1980; BARDEEN, 1971; BOSS, 1980) and analytically (BOSS, 1982; MÖHLMANN, 1984a) that rings are not stable against non-axisymmetric perturbations. Consequently, ring shaped configurations should appear as temporary, intermediate or transient states. Following BODENHEIMER's fragmentation-scenario, a sufficiently fast rotating cloud might fragment therefore through transient ring states into two or more matter accumulations orbiting around each other and transfering on this way the spin of the cloud into orbital angular momentum of the (yet rotating) fragments which might undergo further (hierarchical or cascadic) stages of rotational fragmentation as they continue to collapse to higher densities (BODENHEIMER, 1978; BODENHEIMER and TSCHARNUTER, 1979; NORMAN et al., 1980). It shall be noted here that this scenario of rotational fragmentations quite naturally explains the prevalence of binary and multiple stars. Furthermore, it can be understood on the basis of this scenario that protostars may then evolve through further gravitational collapse, when, after a fragmentation sequence, most of the cloud-spin has been transformed into orbital angular momentum of the fragments, generating on this way slowly rotating fragments which fulfill the JEANS criterion and which may collapse without further essential constraints due to rotation. Therefore, collapse models neglecting rotation should be applicable to describe most of the essential physics of the evolution of the latest slowly rotating fragment towards a protostar. This conclusion is the basis for models of the further protostellar evolution,

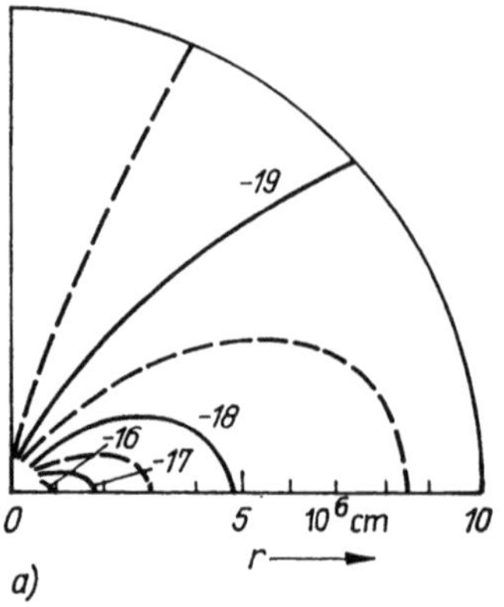

Fig. 5.1 a. Density distribution resulting from a gravitational collapse (from LARSON, 1973). The numbers give the log of the density.

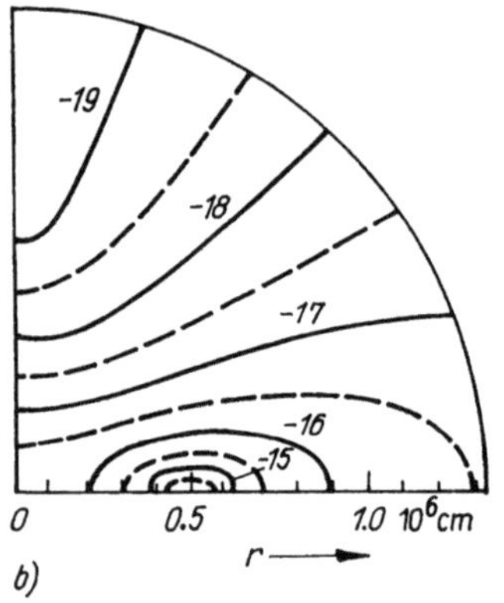

Fig. 5.1 b. An enlargement of Fig. 5.1 a, showing more clearly the ring shaped region of maximum density (from LARSON, 1973). The numbers give the log of the density.

as it will be discussed in section 5.1.5. The main correction due to rotation is that the real collapse process should proceed from a disk instead an isotropic initial state.

Of course, it has to be noted here that the model of rotational fragmentation is not generally accepted as the dominating fragmentation mechanism. This is due to the fact that actions of magnetic and frictional forces or large-scale motions, which were not taken into account, may modify or be an independent cause for the collapse and the fragmentation of interstellar clouds (DORFI, 1981; see section 5.1.4.). It is especially the outward transport of angular momentum by magnetic and frictional forces which may modify the fragmentation process. But these more detailed modifications are not essential for the aim of this section, namely to understand qualitatively the initial conditions for an evolution towards single protostars and their environment, which evolve by a gravitational collapse. Based on the above given discussion, and assuming BODENHEIMER's fragmentation-scenario to be appropriate, these conditions can be characterized als follows:

— collapse preventing spin angular momentum will be transfered (through transient ring phases) into orbital angular momentum,

- at the end of this sequence single protostars may appear whose further collapse (from a disk-like configuration) is not essentially hindered by centrifugal forces. Modifications due to rotation should be mainly restricted to a more flat configuration of the collapsing matter and to a prolongation of the time-scale,
- the remaining relatively low angular momentum is the cause for a later existence of a non-contracting circumprotostellar disk.

The more general and essential results is that the appropriate conditions for the further evolution of protostars by gravitational collapse within a "single-fragment-approach" seem to be relatively independent of the earlier mechanisms of fragmentation and angular momentum transfer, which may proceed on different ways (see Boss, 1985). In every case, and relatively independent of the "fragmentational history", single protostars may evolve when a sufficient amount of local angular momentum of the fragments has been transfered into their orbital momentum.

5.1.4. Interstellar Magnetic Fields

The influence of magnetic fields on the dynamics of interstellar clouds can be estimated from equ. (5.5), indicating in principle, that the action of magnetic fields has to be taken into account even in not collapse-compressed dark interstellar clouds. Possibly, this influence may increase due to flux-conservation in more compressed later stages of the collapse. These conclusions can be justified if the matter in interstellar clouds is sufficiently ionized to be influenced dynamically by the Lorentz-force $\boldsymbol{j} \times \boldsymbol{B}$, which determines the interaction between mechanical and electromagnetic energies. Consequently, the electrical conductivities σ_0, σ_1 and σ_2 in the tensor of electroconductivity

$$\sigma_e = \sigma_0 \boldsymbol{tt} + \sigma_1(I - \boldsymbol{tt}) + \sigma_2 \boldsymbol{t} \times I \tag{5.14}$$

in Ohm's law

$$\boldsymbol{j} = \sigma_e(\boldsymbol{E} + \boldsymbol{v} \times \boldsymbol{B}) \tag{5.15}$$

should not be negligibly small. Here, the following symbols have been used: $\boldsymbol{t} = \boldsymbol{B}/B$ — tangential unit vector parallel to $\boldsymbol{B}$, I — unit tensor, $\boldsymbol{tt}$ — dyadic product, σ_0 — Spitzer conductivity parallel to $\boldsymbol{B}$, σ_1 — Pedersen conductivity, σ_2 — Hall conductivity. For the case that collisions between electrons and ions can be neglected, these conductivities are given by (Möhlmann, 1971)

$$\sigma_0 = \frac{e^2 n_e}{\nu_e m_e} + \frac{e^2 n_e}{\nu_i m_i}, \tag{5.16}$$

$$\sigma_1 = \frac{e^2 n_e}{m_e} \cdot \frac{\nu_e}{\nu_e^2 + \omega_e^2} + \frac{e^2 n_e}{m_i} \cdot \frac{\omega_i}{\nu_i^2 + \omega_i^2}, \tag{5.17}$$

$$\sigma_2 = \frac{e^2 n_e}{m_e} \frac{\omega_e}{\nu_e^2 + \omega_e^2} - \frac{e^2 n_e}{m_i} \frac{\omega_i}{\nu_i^2 + \omega_i^2}, \tag{5.18}$$

where $e = 1.602 \cdot 10^{-19}$ As, m_e and m_i are the mass of electrons and ions, ν_e and ν_i the collision frequencies of electrons and ions with the neutrals, ω_e and ω_i the cyclotron frequencies $\omega = eB/m$ of electrons and ions respectively.

As can be found from equs. (5.1) and (5.5) and with a typical dimension of interstellar clouds of $R \sim 10^2$ pc it follows for the induction of the interstellar magnetic field

$$B \sim 3 \cdot 10^{-10} \frac{\mathrm{V_s}}{\mathrm{m}^2} = 3 \cdot 10^{-6}\ \mathrm{G} \tag{5.19}$$

what agrees well with astronomical results (Mouschovias, 1979).

With $\nu_e/m_e \approx \nu_i/m_i$ and $\nu_e \approx 10^{-4}\ \mathrm{s}^{-1}$ and $n_e \approx 10^8\ \mathrm{m}^{-3}$ (Cap, 1970) follows for the order of magnitude

$$\sigma_0 \approx 10^4 \frac{1}{\Omega\mathrm{m}}; \quad \sigma_1 \approx 10^{-8} \frac{1}{\Omega\mathrm{m}}, \quad \sigma_2 \approx 10^{-2} \frac{1}{\Omega\mathrm{m}}, \tag{5.20}$$

indicating the highly anisotropic character of the action of magnetic fields in interstellar clouds. This anisotropy and the resulting mathematical complications are the main cause for the yet unsatisfactorily low level of knowledge about the detailed role of magnetic fields in interstellar space.

The above mentioned need for sufficiently high electric conductivities can now be formulated more precisely by stating, that for magnetic fields to be important for the dynamics of collapsing interstellar clouds the magnetic dissipation time due to (anisotropic) diffusion out of the collapsing region should be greater than the mechanical time scales for collapse to proceed, or in other words, that the magnetic diffusion-velocity should not exceed the collapse-velocity. This anisotropic diffusion may be described with $\partial \boldsymbol{B}/\partial t = -\mathrm{curl}\, \boldsymbol{E}$ and equ. (5.15) by

$$\frac{\partial B}{\partial t} + \mathrm{curl}\, \sigma^{-1} \mu_0^{-1}\, \mathrm{curl}\, \boldsymbol{B} - \mathrm{curl}\, \boldsymbol{v} \times \boldsymbol{B} = 0. \tag{5.21}$$

The detailed description of this anisotropic diffusion is out of the scope of this book, but for qualitative discussion and estimation it follows for the order of magnitude of the maximum velocity of anisotropic diffusion in the collapsing cloud

$$(v_{\mathrm{diff}})_{\mathrm{max}} = \frac{1}{\mu_0 \sigma_1 R} \approx 10^2\ \mathrm{m/s}, \tag{5.22}$$

what is by orders of magnitude smaller than the collapse-velocity $v \approx (\gamma M/R)^{1/2}$ of a non-fragmentated Jeans-unstable dense interstellar cloud. Consequently, magnetic diffusion (and dissipation) can be neglected in first approximation describing the collapse, or in other words, the dynamics of magnetic fields is directly coupled with that of the interstellar matter, as it can be understood within tho "frozen-flux-approximation".

This intensive coupling between magnetic fields and dynamics of interstellar matter has often been misinterpreted in assuming that the action of magnetic fields is restricted to that of an additional magnetic pressure $p_m = B^2/2\mu_0$, counteracting any contraction as pressure generally does.

It has become clear by the pioneering work of Alfvén, that interstellar magnetic fields may not only counteract but also assist a contraction of interstellar clouds (Alfvén, 1981). This is mainly due to the fact that the action of magnetic fields, as it is described by the Lorentz-force $\boldsymbol{k}_L$, can be formulated as

$$\boldsymbol{k}_L = \boldsymbol{j} \times \boldsymbol{B} = -\mu_0{}^{-1}\boldsymbol{B} \times (\operatorname{curl} \boldsymbol{B}) = -\operatorname{grad} \frac{B^2}{2\mu_0} + \left(\frac{\boldsymbol{B}}{\mu_0} \operatorname{grad}\right) \boldsymbol{B}. \tag{5.23}$$

It is the vector-gradient term in equ. (5.23) which appears additionally to the above mentioned magnetic pressure p_m and which may be essential by pinching the cloud, as has been advocated for a long time by Alfvén. This pinch effect may produce a contraction of a region with initially lower density, being on this way a trigger mechanism for further gravitational collapse (see too Dorfi, 1981).

The filamentary structure of interstellar clouds can be understood as a further result of the action of magnetic fields if they are interpreted as to be connected with force-free magnetic fields (Alfvén, 1981), as it is inferred by the dominance of the Spitzer conductivity $\sigma_0 \gg \sigma_1$, $\sigma_0 \gg \sigma_2$.

Furthermore, magnetic fields are also able to contribute to the solution of the angular momentum problem by their large-scale action which may be the cause for a large-scale angular momentum transport.

In order to get an estimation of the order of magnitude, and if pressure and rotation are neglected, it follows as a remarkable coincidence from equ. (5.1)

$$\frac{3}{5} \gamma \frac{M^2}{R} \approx \frac{4}{3} \pi R^3 \frac{B^2}{\mu_0}, \tag{5.24}$$

what, by assuming that the frozen-flux-approximation may be used throughout the collapse, gives for the magnetic induction concentrated in the protostellar environment ($M \sim 1 M_\odot$, $R \sim 3 \cdot 10^{12}$ m)

$$B \approx \left(\frac{9}{20\pi} \gamma \mu_0\right)^{1/2} \frac{M}{R^2} \approx 3 \cdot 10^{-4} \frac{v_s}{m^2} = 3G. \tag{5.25}$$

This corresponds quite well to the order of magnitude of magnetic fields, assumed to have been present during the formation of solid matter of some primitive (and consequently early formed meteorites (Brecher, 1972).

All these estimations strengthen the conclusion that interstellar magnetic fields have been present and non-negligible in star forming collapse processes. But, and as can be seen from the above given discussion, the involvement of the Lorentz-force complicates quite essentially the complicated mathematical treatment of collapse processes. A first successful step into this direction has been done by Dorfi (1981) with his three-dimensional calculations on the dynamics of rotating, magnetized and self-gravitating interstellar clouds. In Dorfi's calculations the magnetic field is treated as "frozen in", and a homogeneous induction of $3 \cdot 10^{-6}$ G is assumed to be either perpendicular or to be aligned with the axis of rotation. The main results are:

— The magnetic field may prevent the major part of the cloud from collapsing and favours the formation of highly nonhomologous, barlike structures.
— The magnetic field is able to transfer angular momentum at timescales that are short compared with the time of free fall.

Summarizing it has to be stated that

a) the action of interstellar magnetic fields can not be neglected in detailed models even of the early phases or the triggering of gravitational collapses,
b) the necessary inclusion of magnetic fields in the models of the early stages of star formation essentially increases the possible variety of evolutionary ways,
c) general consequences of the action of magnetic fields in prestellar interstellar clouds are
   — possible transfer of angular momentum,
   — assistance to contractive motions with resulting locally enhanced densities, which may result in condensation and growth of grains, and which may trigger a further gravitational collapse,
   — the evolution of more linear, filamentary or bar-like structures in interstellar matter what might give quite another fragmentation-sequence towards star formation than that from more or less isotropic interstellar cloud fragmentation by the BODENHEIMER scenario. It is interesting to note here, that even in 1930 NÖLKE postulated the evolution of the Solar system from a bar-like structure. For a more detailed description of the above discussed fragmentation processes, see BOSS (1985).

5.1.5. Evolution towards and of Protostars

It has been suggested in sections 5.1.2. and 5.1.3. that the description of the evolution towards a protostar can be based on the three-dimensional model of a gravitational collapse of a slowly rotating fragment with a mass exceeding the JEANS mass. The neglection of rotation in a first approximation gives the possibility to restrict the computations on a spherically collapsing fragment. The derived results do not deviate essentially for the more realistic model of the formation of protostars from a flat (disk-like) distribution of matter.

A further aspect to use the simpler isotropic models is that the spherical case is up to now the only "three-dimensional" one in which it has been possible to follow the collapse from the fragment to the formation of a hydrostatic pre-main sequence star. This task has been successfully attacked first by LARSON (1969), using some simplifying assumptions, which have been questioned by NARITA et al. (1970) but justified by consistent model calculations of WINKLER and NEWMAN (1980) without any additional assumptions. Also the work of RÓZYCZKA et al. (1980) and TSCHARNUTER (1985) should be noted here.

In general, it can be stated, that this work has not been finished and also unexpected solutions as transient oscillations (TSCHARNUTER, 1985) of the forming protostellar core can not be excluded, but nevertheless, there is consense in the main steps of protostellar evolution. They shall be discussed in the following, based on LARSON's classical work in this field. This seems to be sufficient for the aim of this book. The onliest deviation which may be essential is that a (rotationally caused) more flat pre-protostellar disk can be a more appropriated initial condition.

The model parameters for the fragment used by LARSON (1969, 1972) are given by Table 5.3. They are nearly identical with those of WINKLER and NEWMAN (1980).

Table 5.3. Initial and boundary conditions of collaps models

M	$= 1\ M_0$
T	$= 10$ K
R	$= 1.63 \cdot 10^{15}$ m
ϱ_0	$= 1.1 \cdot 10^{16}$ kg m^{-3}
k/μm	$= 3.36 \cdot 10^3$ Ws/kg K

This model corresponds to a composition of 65% molecular hydrogen, 32% helium and 3% heavier elements of the fragment (GAUSTAD, 1963).

An essential property of the gravitational collapse is the free fall time t_{ff}, whose order of magnitude can be estimated from $mv^2/2 = \gamma Mm/R$ and $v - (2\gamma M/R)^{1/2}$ by

$$|t_{ff}| = \int_0^R \frac{dr}{v(r)} = (6\pi\gamma\varrho)^{-1/2}, \tag{5.26}$$

what differs by a factor of the order of unity from similar representations, derived under somewhat different assumptions.

Obviously, this free fall time has to be smaller than the characteristic time for the propagation of thermal effects, $t_{th} = R/c$, where $c^2 = \bar{\gamma}kT/\mu m_H$ is the velocity of sound. Otherwise, any gravitational density enhancement would be dispersed by the action of pressure. By the way, and as can be shown with $R = (3M/4\pi\varrho)^{1/3}$, the relation $t_{ff} = t_{th}$ is equivalent to the equilibrium condition, formulated with the JEANS criterion.

As has been mentioned in section 5.1.3., the fact that the free fall time is shortest where the density is highest, is the cause for the non-homologous character of the collapse. This deviation from a non-uniform collapse is of great importance for the evolution of the collapse, since the major changes in density occur increasingly faster in an increasingly smaller region at the center and on a shorter and shorter time scale, while almost nothing happens in the outer part of the cloud, which still contains most of the mass (LARSON, 1978). A further remarkable property of the collapse has been mentioned in section 5.1.1. It can be seen from Table (5.1) that the collaps is isothermal in its earlier stages. The resulting density distribution for the isothermal spherical collapse can be shown to be asymptotically (for large distances from the center) of the form (LARSON, 1973)

$$\varrho = 0.705 \frac{kT}{\gamma \mu m_{\mathrm{H}}} r^{-2} . \tag{5.27}$$

In a similar way the velocity distribution approaches

$$v = 3.28 \left(\frac{kT}{\mu m_{\mathrm{H}}} \right)^{1/2} , \tag{5.28}$$

indicating that the collapse velocity during the isothermal phase never exceeds 3.28 times isothermal sound speed (LARSON, 1973).

Furthermore, the numerical computations of LARSON (1973) led to the result, that at the center of the cloud the ratio of pressure to gravity forces is 0.6, so that the acceleration of the collapse is reduced to 0.4 times the free fall acceleration and the free fall time increases by a factor of 1.58. A further reduction may be supposed as to be caused by centrifugal forces. At large distances from the center, the ratio of pressure to gravity forces drops to a limiting value of 0.23 (LARSON, 1973).

A qualitative change occurs in the character of the collapse when the central density reaches a value of the order of 10^{-10} kg m^{-3}. Then the central part of the cloud starts to become opaque also for the infrared radiation from the dust grains. Consequently, the thermal energy generated by the collapse can not be radiated away and the central temperature starts to rise significantly over the initial value of about 10 K. The further collapse of the center now becomes nearly adiabatic rather than isothermal and the thermal pressure begins to rise rapidly in the center. When the central density reaches $\sim 10^{-9}$ kg m^{-3} the rising pressure effectively decelerates the collapse at the center. When the density is of the order of some 10^{-7} kg m^{-3}, the central collapse will be stopped. The temperature at this time is close to $2 \cdot 10^2$ K (LARSON, 1973), the mass and radius of the core are approximately $6 \cdot 10^{-3} M_\odot$ and $7.5 \cdot 10^{11}$ m respectively. About 300 years later the central density has reached 10^{-4} kg m^{-3}; the core mass and radius are about $1.1 \cdot 10^{-2} M_0$ and $2.5 \cdot 10^{11}$ m respectively and the temperature increased to 1900 at the core. Now most of the internal energy generated by the collapse goes into dissociating the hydrogen molecules, the core becomes isothermal again and collapses non-homologously towards a sharp central peak in the density distribution. When nearly all of the hydrogen molecules have been dissociated, central temperature and pressure begin again to increase and to stop the collapse, forming a second hydrostatic core at the center of the first core. The material of the first core falls towards the centre and accretes on the second core within a time of ~ 10 years. When the second core forms up, it has a mass of about $1.5 \cdot 10^{-3} M_\odot$, a radius of $\sim 1.2 R_\odot$, a central density of $\sim 2 \cdot 10^1$ kg m^3 and a central temperature of $\sim 2 \cdot 10^4$ K (LARSON, 1973). The time elapsed from the beginning of the collapse to the formation of the second core is one or two times the free fall times, i.e. of the order of about $3 \cdot 10^5$ Years (it is of the order of the "Helmholtz"-time for a gravitational collapse).

Calculations of the stages of evolution following the formation of the second core (protostellar core) have been carried out by LARSON (1969, 1972). The protostellar core reaches its maximum radius about 1 year after its formation, having then a mass of $\sim 10^{-2} M_\odot$ and a radius of $\sim 15 R_\odot$. Nearly all of the energy emitted from the matter accreting core is effectively radiated away, although it does not immediately leave the circumprotostellar environment, but is first absorbed by the dust grains in the cooler outer regions of this cloud, from which it is reemitted thermally at infrared wavelengths.

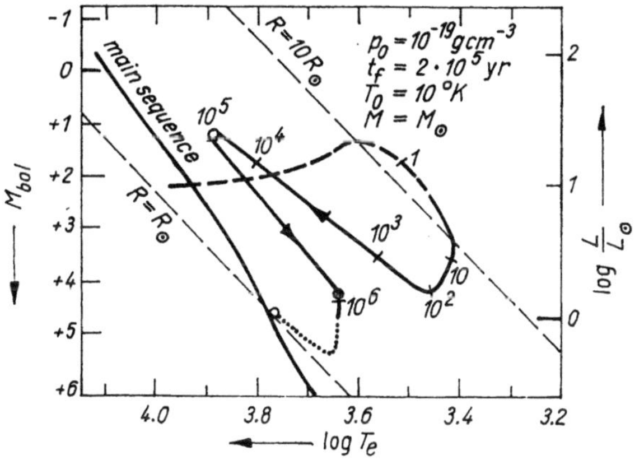

Fig. 5.2. The evolution of the stellar core in the HR diagram for a protostar of mass 1 $M_\odot$. The numbers marked along the track give the time in years since the formation of the stellar core (from LARSON, 1973).

The further evolution of the protostellar core is illustrated in Fig. 5.2. At a time of about 100 years after the formation of the protostellar core the core radius is about $5 R_\odot$ and its mass is still close to $10^{-2} M_\odot$. The next phase between 10^2 years and 10^5 years has to be characterized by a decrease of the radius but an accretional increase in the core mass. Especially, about $7 \cdot 10^4$ years after the core formation about half of the total mass has been accreted to the protostellar core, the radius of the core is then $\sim 3 R_\odot$, the effective temperature is of $\sim 8 \cdot 10^3$ K and the luminosity is about $25 L_\odot$. As will be shown later, even at this early phase planetogonic processes can be working. At this time a convection zone evolves at the surface of the core and begins to grow inward, taking place then deep in the interior of the core, while the core contracts somewhat as it cools down. This might be of importance for early dynamo-generated magnetic fields. Simultaneously the kinetic energy inflow by accreting matter continues to decrease as the infalling environment becomes depleted of material. The time elapsed since the formation of the core is now about $9 \cdot 10^5$ years; the time elapsed since the beginning of the collapse is about $1.2 \cdot 10^6$ years. The protostar should become observable as a T Tauri star.

At this time the energy inflow due to accretion has become unimportant compared with the radiative energy outflow from the stellar surface which results from the gravitational energy, released by the further contraction of the core. The resulting luminosity can be described by

$$L = 4\pi^2 R^2 \sigma T_{\text{eff}}^4 = \frac{\mathrm{d}\,E_G}{\mathrm{d}t} = \frac{3}{5}\gamma\,\frac{M^2}{R^2}\,\frac{\mathrm{d}R}{\mathrm{d}t}, \tag{5.29}$$

having the solution

$$R = t^{-1/3}\left(\frac{\gamma M^2}{20\pi\sigma T_{\text{eff}}^4}\right)^{1/3}. \tag{5.30}$$

On the other side, as a time-scale for this so-called "Kelvin-Helmholtz" contraction follows

$$t_{\text{KH}} \approx \frac{\gamma M^2}{20\pi\sigma T^4 R^3}. \tag{5.31}$$

As a remarkable matter of fact, a decrease of the radius of T-Tauri stars with increasing time can be predicted from these estimations. This is in agreement with astronomical observations, as will be shown in section 5.1.6.

When the protostar reaches its T-Tauri pre-main sequence stage, it has a radius of about $2R_\odot$, an effective temperature of 4400 K and a luminosity of about $1.5L_\odot$ (Larson, 1972).

The density distribution in the steadily infalling cloud can be derived from $4\pi r^2 \varrho v =$ const. and $v = (2\gamma M/R)^{1/2}$ to be of the form

$$\varrho = \varrho_0 r^{-3/2}. \tag{5.32}$$

The outermost part of the protostellar environment has still the "isothermal" density distribution, as it has been described by equ. (5.27). Of course, these profils may be changed slightly especially due to the action of centrifugal forces and possible the redistributing action of viscous forces and magnetic fields. Especially the profile in the outer parts should be steepened due to the later radial outflow of matter from inside (as a result of angular momentum conservation and thermal effects).

It has to be noted here, that the accuracy of the numerical values given in this section should not be overestimated. This is due to their dependence on the initial conditions. If for example the radius of the fragment is smaller by a factor of 10 as it has been used above, radius and luminosity of the resulting star are increased to about $6R_\odot$ and $9L_\odot$ and the total formation time is reduced to about $3.5 \cdot 10^4$ years. As a limiting case, Hayashi and coworkers (see Hayashi, 1966; Narita et al., 1970) started their computations for a $1M_\odot$ protostar model with a radius of only $5 \cdot 10^3 R_\odot$, a polytropic sphere of index 4 and a resulting central density of $7 \cdot 10^{-6}\,\text{kg m}^{-3}$ (corresponding to a mean density of $10^{-8}\,\text{kg m}^{-3}$, what is higher by eight orders of magnitude then Larson's initial density). These initial conditions have been based on the assumption, that a fragment collapsing to sufficiently high densities becomes completely opaque and its structure can be approximated by a polytropic sphere. As has been discussed in section 5.1.3. and in this section, it is the strongly non-homologous character of the collapse, causing the fact that only a very small region at the center becomes opaque and most of the mass remains in the extended optically thin outer regions of lower density. However the calculations of HAYASHI and coworkers can be understood as a limiting case for situations where a collapsing precompressed fragment may be caused by external pressure from supernovae or spiral density waves.

It is obvious that for an initial condition of a flat preprotostellar configuration the above given numerical results change too.

Summarizing the results of the above reviewed calculations, it can be stated, that the evolution of low mass protostars and their circumprotostellar environments may be characterized by the following conclusions:

1. The degree to which the collapse is non-homologous and the relative amount of angular momentum determines essentially the resulting final states.
2. If due to external forces (compression) the protostellar environment is extremely dense, what would correspond to a weak non-homologous collapse, the infalling material may become so opaque that no significant radiative energy losses occur. As a consequence of the resulting heating of the protostellar matter a $1M_\odot$ protostar may attain a large radius of $\sim 100R_\odot$ and a corresponding high luminosity of $\sim 500L_\odot$ (NARITA et al., 1970; LARSON, 1973).
   More generally it can be stated that in cases of rapid accretion (weakly non-homologous collapse) the Kelvin-Helmholtz time quickly becomes of the order of the accretion time. As can be seen from equ. (5.26), this is connected with a relatively large radius and with high luminosity of the protostar. This may essentially influence the thermal character of the remaining circumstellar shell and envelope.
3. If on the other hand the early stages of the collapse are strongly non-homologous, a much more diffuse and extended (and only slowly accreting) region may evolve. This region may be sufficiently optical thin at infrared wavelengths, causing a relatively efficient radiative cooling of the circumprotostellar matter. The KELVIN-HELMHOLTZ contraction time of the still matter-accreting protostellar core increases slowly. This is then connected with a core of a relatively small radius and low luminosity.
4. The final properties of the remaining circumprotostellar envelope depend on the initial density (spherical or flat structure, radial profile) as well as on the degree to which the collapse is non-homologous, what is also dependent on the thermal properties of the collapse and the resulting pressure-caused retardation of the inward motion of the cooler outer regions of the collapsing envelope. Furthermore, the conservation of the total angular momentum of the original fragment will be the cause for the later existence of a stably orbiting planetary or satellite system. Extension and mass of the envelope depend on the combined action of the non-homologous development of the collapse (giving relatively thin or massive envelopes) and the mechanism depositing angular momentum (and connected matter) in outer regions.
5. The collapse-caused increase in the density and the thermal conditions in the

circumprotostellar shell and envelope, connected with changes in outward (radiative) energy transport and protostellar luminosity, may give appropriated conditions for heating and cooling and resulting condensation and reheating processes, contributing on this way to the later preplanetary solid matter.

6. More quantitatively, the following "model-parameters" for the yet evolving circumprotostellar cloud can be used for the calculations in chapter 5.3.:

a) The outermost regions are characterized by low temperatures (of the order of some, 10 K) due to the possibility for infrared radiation to escape (densities not exceeding $(10^{-9}-10^{-10})$ kg/m³).

b) The inner regions should be heated, having higher temperatures than in the case that radiation could escape freely ($T \sim r^{-1/2}$). The resulting temperature profile can be approximated by $T \sim r^{-1}$.

5.1.6. Astronomical Results and Parameters of Circumprotostellar Shells

During most of its evolution, the protostellar core remains heavily obscured by the dust in the optically thick infalling environment. Consequently, the protostar and its environment become observable only at infrared wavelengths. The detailed structure of this circumprotostellar matter has an essential influence on the observable properties of protostars. This is due to the fact, that until later phases, when the protostellar environment is partially dispersed and partially accumulated in a protostellar disk, the luminosity (generated by accretion of the matter at the protostellar surface, or originating inside the core) is absorbed by dust grains in the envelope and converted into infrared radiation. Consequently, the emitted spectrum of a protostar depends essentially on the transfer and propagation of infrared radiation through the envelope. Additionally, it is difficult even to predict with quantitative accuracy the properties of the infrared spectrum of protostars. Probably up to now, only qualitative properties and results can be of significance. This is mainly due to the restriction of computations and models to spherically symmetric conditions, neglecting the modifications resulting from a flattned geometry, and due to the yet important uncertainties in the dust-to-gas ratio and the only purely known infrared optical properties of the dust grains.

This qualitative agreement has indeed been found by observational proof, namely by the fact that collapse motions exist in dense interstellar clouds (Lequeux, 1977; Loren, 1977) and by the qualitative agreement between the predicted spectra and the observed infrared properties of protostar. These observed spectra are generally broader than a black body spectrum. They show absorption features at $3 \cdot 10^{-6}$ m and $10 \cdot 10^{-6}$ m. Finn and Simon (1977) were able to explain the 10 μm feature by a model with silicate grains in the protostellar envelope, assuming a radial power-law density profil of the silicates. An even better approximation to the observed infrared spectra has been found by Cohen (1973) with a double shell model. The assumed "double cocoon" structure is in fact a consequence of models of Yorke and Krügel

(1977), which indicate the existence of an outer shell of ice grains at a temperature of < 200 K and an inner shell of refractory grains at a temperature of ~ 1000 K. This might be of importance as well as for planetogonic processes, which depend sensitively on the structure of the disk as for the chemical constitution of the preplanetary disk. Furthermore, the results of theoretical estimations, as they were given at the end of section 5.1.5. are confirmed by these results.

A further agreement can be found for evolutionary changes of properties of the protostellar spectra. As the protostar evolves, the density of the infalling envelope decreases and the photosphere of the protostar moves inward to higher temperatures. This causes a steady shifting of the peak of the infrared spectrum towards shorter wavelengths. The duration of this infrared stage of evolution is of the order of $5 \cdot 10^5$ years to 10^6 years. The resulting evolutionary tracks in an infrared HR diagram have been computed and compared with observational results by LARSON (1972). The results are shown in Fig. 5.3. With its further evolution, the protostellar shell becomes optically thin. A double-peak spectrum should become visible with contributions from the envelope as well as from the core. These predicted spectra correspond qualitatively to those of a number of T-Tauri stars.

The "T-Tauri" stars were first recognized and defined by JOY (1945) (see also BERTOUT, 1982). Their classification criteria have been given in more precise form by HERBIG (1962). The assumption that T-Tauri stars are low-mass, pre-main sequence objects has been advocated by AMBARTSUMIAN (1947), HERBIG (1977) and JONES and HERBIG (1979) and others.

The optical properties of T-Tauri stars have been reviewed by COHEN and KUHI (1979). Balmer lines are often complex and highly variable, indicating activities at the stars and especially in their envelope. MENDOZA (1966, 1968)

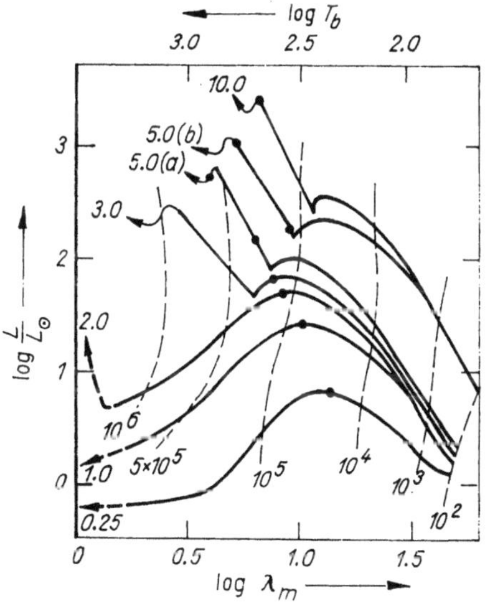

Fig. 5.3. Evolutionary tracks of protostars in an infrared HR diagram (λ_m is the wavelength of the peak IR emission). The protostellar masses from 0.25 $M_\odot$ to 10 $M_\odot$ are marked on the curves. The dashed lines are isochrons for the times marked at the bottom (from LARSON, 1973).

disovered that T-Tauri stars show a large infrared excess, what has been attributed to the cooler dust in the circumstellar envelope. VARDANIAN (1964) discovered that the polarization of some T-Tauri stars is variable. This is caused by the presence of dust in the dynamically active circumstellar matter. Using IUE, some T-Tauri stars have been observed in the UV (APPENZELLER et al. 1980; GAHM, 1980). The emission measure of the temperature distribution in the transition region ($2 \cdot 10^4$ K $< T < 2 \cdot 10^5$ K) is much larger than in the Sun or than in active late stars (BERTOUT, 1982). The continuum radiation at UV wavelengths is probabely due to Balmer emission (BERTOUT, 1982).

In the X-range, T-Tauri stars seem to be very active (FEIGELSON, 1983), what may be caused by strong flares ("superflares"). This has been discussed too in part 3 of this book. Also in the radio range T-Tauri stars have been observed (BERTOUT, 1982). Most of the T-Tauri stars have relatively low rotational velocities (VOGEL and KUHI, 1981).

A yet open problem is the reason for T-Tauri-activity and wether the observed spectral properties have to be explained by the effects of infall from the circumstellar envelope, what seems to be the case with so-called YY Orionis stars with infall velocities between 150 km/s to 400 km/s (WALKER, 1972), and as it has been proposed by LYNDEN-BELL and PRINGLE (1974), or if an outflow of matter, possibly connected with flares or a local "break-trough" of the hot shell trough the surrounding cooler matter, possibly caused by the action of a strong stellar wind, is reponsible for the observed features (ULRICH, 1976).

In order to refer briefly to other stellar types assumed to be connected with young stars, the Herbig-Haro objects shall be noted (HERBIG, 1960; STROM et al., 1972) which seem to be the more massive ($2M_\odot$ to $6M_\odot$) analogues to the "normal" T-Tauri stars, and which are still surrounded by a dense circumstellar envelope. A further interesting phenomenon is connected with the so-called FU Orionis stars which have been identified as T-Tauri stars, but which flare up by about 6 magnitudes within the range of about 100 days and one year. Up to now, there is no generally accepted reason known for this, but it should be mentioned that the collapse models of HAYASHI and coworkers (HAYASHI, 1966) seem to indicate that this flare up may be interpretable as a last stage of the dynamical evolution before quasi-hydrostatic contraction towards the main-sequence occurs.

Further information about the processes in circumstellar matter around newly born stars has been found from advanced infrared — and radioastronomy. A remarkable result is, that regions of young stars are often connected with masers. These H_2O-, OH-, Si- and CH_3OH-masers are concentrated in "active regions" with characteristic diameters of the order of 10^{16} m. The masses of these maser-knots seem to be of the order of planetary masses, and their dynamics seem to indicate the existence of two different groups. Usually they move with the velocity which is typical for their original dense interstellar cloud (of the order of 10 km/s), while a second group shows velocities up to some 10^2 km/s, being directed away from the center of activity. These centers have characteristical lengths of some 10^{14} m. The observations of these maser sources indicate the existence of sharp temperature gradients between cold interstellar matter and hot plasma (of the stellar wind or from H II regions,

Table 5.4. Properties of circumstellar shells of some T-Tauri stars (from GÖTZE, 1980)

Star	Spectr.	T_{eff} (10^3 K)	R_s (10^9 m)	R_s/R_0	T_H (10^4 K)	n_e (10^{16} m^{-3})	R_H (10^{10} m)	R_H/R_s	L_H (10^{26} W)	L_H/L_s	B (10^{-2} Vs/m^2)
BP Tau	K5V	4.40	1.52	3.76	2.00	2.04	2.06	13.53	4.74	0.78	1.00
	K6V	4.20	1.59	4.05	1.50	1.37	2.24	14.11	2.68	0.48	0.79
DF Tau	K6V	4.20	2.48	6.31	2.00	2.04	2.85	11.50	12.7	0.93	1.00
DG Tau	G0V	5.90	0.77	1.03	2.00	2.04	2.36	30.57	7.20	1.40	1.00
	G8V	5.44	0.63	1.18	2.00	2.04	1.81	28.36	3.22	1.33	1.00
DK Tau	M0V	3.75	1.73	4.60	3.00	3.54	1.02	5.91	1.60	0.37	1.58
	K7V	4.00	1.88	4.91	2.00	2.04	1.62	8.62	2.32	0.36	1.00
CO Ori	G5V	5.66	2.32	4.22	0.60	0.30	13.5	58.03	35.2	0.90	0.25
	G2V	5.77	2.32	3.55	0.60	0.30	13.8	59.65	38.3	0.89	0.25
	F8V	6.00	2.21	2.94	0.50	0.16	21.3	96.47	65.6	1.46	0.16
GW Ori	K3V	4.77	3.26	7.00	2.00	2.04	3.14	9.63	15.9	0.41	1.00
	K4V	4.59	3.44	7.38	2.00	2.04	3.00	8.72	13.9	0.37	1.00
IP Mon	K3V	4.77	1.75	3.76	0.50	0.16	12.8	73.02	14.1	1.25	0.16
LU Mon	K5V	4.40	1.26	3.12	1.00	0.72	2.87	22.78	1.80	0.43	0.50
MO Mon	K4V	4.59	1.93	4.14	1.00	0.72	4.54	23.52	7.14	0.61	0.50
MM Mon	K3V	4.77	1.38	2.96	1.00	0.72	3.89	28.17	4.49	0.64	0.50
LH 188	K5V	4.40	1.85	4.58	0.60	0.30	5.88	31.79	2.93	0.33	0.25
LH 191	K2V	4.96	1.65	3.49	0.50	0.16	10.5	63.90	7.93	0.67	0.16

(T_{eff} – effective temperature, R_s – stellar radius, R_0 – radius of a normal main sequence star of the same spectral type, T_H – temperature of the circumstellar shell, n_e – electron density in the centre of the shell, R_H – radius of the shell, L_H – luminosity of the shell, L_H/L_s – ratio of the luminosities of the shell and the star, B – a well founded estimation of the amplitude of the magnetic induction.)

generated by the young stars). These maser sources are interpreted therefore as to be connected with local accumulations of dust within the above mentioned expanding cocoons (Bettwieser, 1978). The further evolution (and the planetogonic consequencies) of these small and local mass accumulations after their maser-phase is yet unknown. Probably they may act as further (but smaller) collapse centers, which may escape from their original protostar, or may be only slightly bound gravitationally in great distances (of the order of 10^{16} m) from their protostar. A further yet open question is whether masers are typical only for massive ($M > M_\odot$) protostars. The physical parameters, characterizing circumstellar shells of T-Tauri stars have been investigated by Götz (1984).

The results shall be summarized in Table 5.4.

Furthermore, it has been stated by Götz (1984) that the relatively high shell temperatures and electron densities are connected with relatively small shell diameters and relatively large stellar radii. Later stages have a more extended shell of lower temperature and electron density and smaller stellar radii. A further result is that the rotational angular velocity seems to decrease while the T-Tauri stars approach the main sequence. Furthermore, it should be repeated that the infrared excess of the T-Tauri stars is not caused by the hot circumstellar shells but by the circumstellar dust outside of these hot shells in the circumprotostellar envelope. Possibly, condensation of circumstellar matter, flowing out of the hot shells, may contribute on this way to the formation of the dust out of the shell (Dorschner, 1971).

Further information, particularily on the influence of rotation on the geometry of the matter in circumstellar envelopes around newly formed stars has been found from investigations of the star formation region of S 106. In spite of the fact, that due to greater mass of the central object and greater extension of the circumstellar matter of S 106 a quantitative comparison can not be made with the protostellar environment of low mass stars and the corresponding extensions in early stages of their protostellar collapse, the properties of the star formation region S 106 shall be reviewed, here, to give some qualitative characteristics of the environment of newly formed stars. According to Gehrż et al. (1982) S 106 is almost certainly a region of star formation which is associated with a massive $2 \cdot 10^4 M_\odot$ dense interstellar cloud (molecular cloud), an H_2O- and an OH-maser. The general morphology of this region has been illustrated by Fig. 1.1. It can be characterized as a bionical nebula, inclined by about 30° to NW with the northern and southern lobes separated by a lane of dust (Eiora et al. 1979). Eight compact infrared sources have been identified. Their spectra are illustrated in Fig. 5.4. The interpretation, given by Gehrz et al. (1982) states, that

- the IRS 4 is exiting the entire optical, infrared and radio structure with a diameter of $\sim 10^4$ AU, it is a 07—09 star (Panagia, 1973),
- each of the lobes has a diameter of ~ 0.1 pc,
- the total luminosity amounts to about $(2-3) \cdot 10^4 L_\odot$,
- the near infrared radiation may emanate from an optically thick toroidal volume of $R \sim 24$ AU which is heated to ~ 900 K by the central star (which is more massive than the Sun),

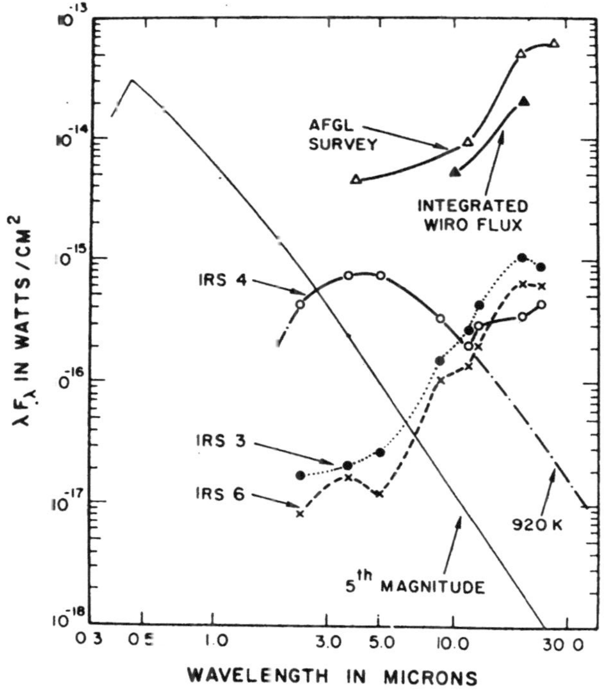

Fig. 5.4a. Broadband IR spectra for selected compact infrared sources associated with S 106. This Figure indicates the short wavelength dominance of IRS 4 whose color temperature is 920 K (from GEHRZ et al., 1982).

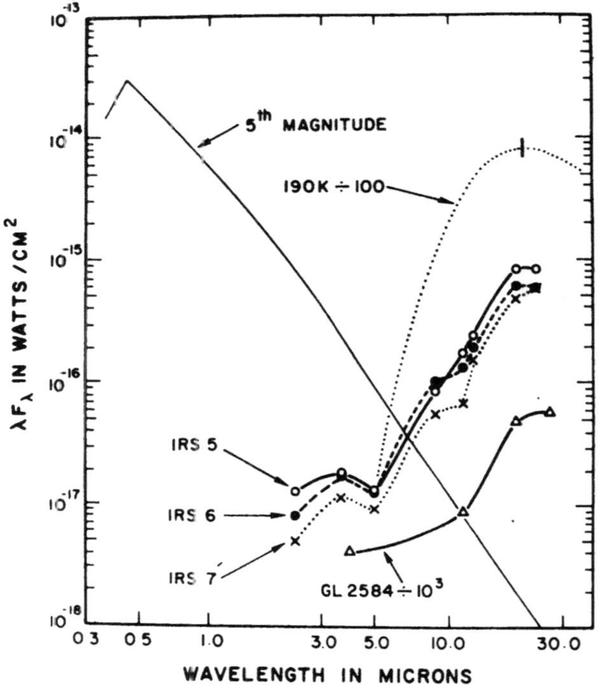

Fig. 5.4b. Broadband IR spectra for selected compact infrared sources associated with S 106. The dotted line labeled 190—100 shows the flux expected from a 190 K blackbody source. This is the approximate color temperatur of the cool sources in S 106 (from GEHRZ et al., 1982).

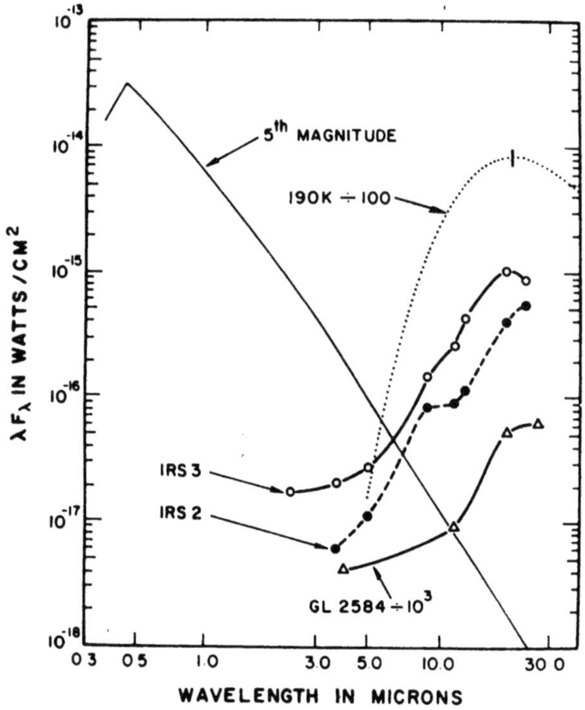

Fig. 5.4c. Broadband IR spectra for selected compact infrared sources associated with S 106. The color temperatures of IRS 2 and IRS 3 are of $\sim$ 170 K. There is no a 3.6 μm excess emission as in Fig. 5.4.3. (from GEHRZ et al., 1982).

— some of the compact infrared sources are not additional luminous exciting stars but might well be the sites of formation of low luminosity companions to the central source,
— rotation has played on essential role in determining the geometry of S 106,
— in the north and south lobes there is evidence for material flowing outwards at velocities as high as 100 km/s.

A schematical scetch of some properties of S 106 is given by Fig. 5.5, relating this possibly to structures resulting from the action of an effective magnetic field of the central body. This is discussed more in detail in section 5.2.2. and with Fig. 5.8.

A further interesting observational result comes from IRAS-observations of dust disks around Wega and other stars. But very probably, these indications for grains accumulated in a far disk above the evolved star are not connected with planetogonic processes. They could be the result of an outward transport of grains by radiation pressure.

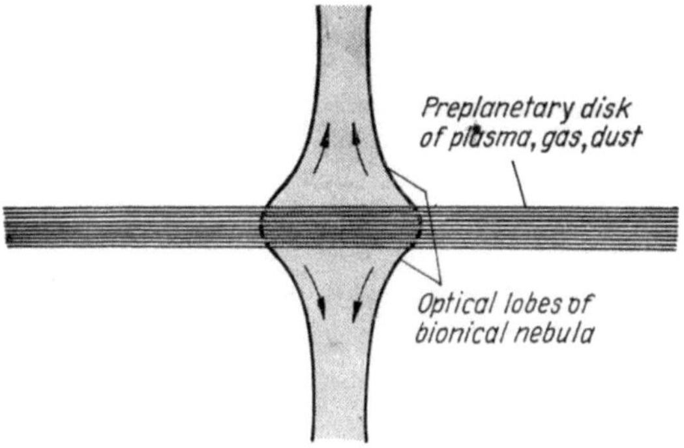

Fig. 5.5. Schematical sketch of the Alfvén surface (for partial corotation) as the envelope for the lobes of biconical nebula. Outside the Alfvén surface the matter moves towards the equatorial plane, forming there thin preplanetary disk.

Summarizing the planetogonically relevant results of chapter 5.1. it can be stated that:

- protostars form from fragments of interstellar clouds with initial densities of about 10^{-17} kg m^{-3} and temperatures of about 10 K,
- the properties of the remaining circumprotostellar environment depend strongly on the degree to which the collapse is non-homologous and on the involved total amount of angular momentum of the original fragment,
- for the case of a strongly non-homologous collapse a hot inner circumprotostellar environment evolves while a more diffuse and extended (and only slowly accreting) outer region with densities of $(10^{-9}-10^{-10})$ kg m^{-3} remains,
- when the mass of the growing protostar becomes comparable to that of the environment (disk) the Jeans mass reaches planetary masses. As will be described later (in chapter 5.3.) there are now appropriated conditions for the formation of large protoplanets (via transient ring-phases which decay due to an azimutal instability),
- simultaneously, the growing protostar should be able to generate large-scale magnetic fields (due to the envolving convection zone). This magnetic field could be able to modify physical state and conditions in the inner (hot) part of the circumprotostellar disk,
- at the same time even in this (pre-)T-Tauri-phase, conditions are appropriated for the formation of solid matter (dust, larger blocks). This is due to the fact, that the scales for local collective gravitational instabilities decrease with time, leaving at the end of this phase (T-Tauri and later) only the possibility for the formation of km-sized bodies via the "GOLDREICH-WARD-instability (see section 5.3.8.).

5.2. Evolution towards Preplanetary Disks

The physical processes in the protostellar environment are characterized by complex interactions of gravitation and magnetic fields from the protostar with the forming disk of interacting gas, plasma and dust. It is the aim of this chapter to describe physical processes which seem to be of relevance for the evolution of this protostellar environment towards a thin disk, which according to the proposed planetogonic scenario will be structurated into rings which are the predecessors of planetary bodies.

The importance of the action of early magnetic fields for the evolution of the preplanetary disk is described in 5.2.2. The inner boundary for the early preplanetary disk, which can be a result of the action of magnetic fields from the protostar (protoplanet), can be connected with the existence of "primary" rings, as it has been discussed in chapter 5.2.3., and with the formation of a relatively mass-rich disk (due to the damming up of the inflow of matter).

The influence and results of neutral gas processes (as pressure, turbulence and friction) and of gas-dust interactions (growth of grains, radial transport, sedimentation in the disk) are investigated in the following chapters.

The action of these processes, including that of the radiation pressure, leads at least to the formation of an inner disk, consisting mainly of particles and gas, and an outer and cooler and more gaseous disk.

5.2.1. Basic Equations

The coupled equations of mechanics and electrodynamics describing properties of circumstellar matter of plasma, gas and dust are given in the (magneto-) hydrodynamic approximation for each of these species by the corresponding balance equations for mass density

$$\frac{\partial}{\partial t}\varrho + \operatorname{div}\varrho\boldsymbol{v} = 0, \tag{5.33}$$

momentum density

$$\frac{\partial}{\partial t}\varrho\boldsymbol{v} + (\boldsymbol{v}\nabla)\,\varrho\boldsymbol{v} = \boldsymbol{k} \tag{5.34}$$

with the force density

$$\boldsymbol{k} = -\operatorname{grad} p - \varrho\operatorname{grad}(V_0 + \phi) + \operatorname{div}\mathbb{P} + \boldsymbol{j}\times\boldsymbol{B}, \tag{5.35}$$

the involved stress tensor

$$\mathbb{P} = \eta(\nabla\circ\boldsymbol{v} + \boldsymbol{v}\circ\overleftarrow{\nabla}) + I(\eta' - 2\eta)\operatorname{div}\boldsymbol{v} \tag{5.36}$$

and Ohm's law

$$\boldsymbol{j} = \sigma(\boldsymbol{E} + \boldsymbol{v}\times\boldsymbol{B}) \tag{5.37}$$

and for the angular momentum density $\boldsymbol{l} = \varrho \boldsymbol{r} \times \boldsymbol{v}$

$$\left(\frac{\partial}{\partial t} + \nabla \boldsymbol{v} \circ\right) \boldsymbol{l} = \boldsymbol{r} \times \boldsymbol{k}. \tag{5.38}$$

This set of equations has to be closed by Poisson's equation $\Delta\phi = 4\pi\gamma\varrho$ for the gravitational potential ϕ of circumprotostellar matter, an appropriate equation of state $p = p(\varrho, T)$, and by Maxwell's equations, giving in the non-relativistic case $v \ll c$ for the Lorentz-force the representation

$$\boldsymbol{j} \times \boldsymbol{B} = -\nabla\left(\frac{\boldsymbol{B}^2}{2\mu_0} I - \frac{\boldsymbol{B} \circ \boldsymbol{B}}{\mu_0}\right). \tag{5.39}$$

The force density $\boldsymbol{k}$ has to be modified, if special interactions, such as collisions between the neutrals and dust have to be taken into account.

Here the usual symbols have been used:

$\varrho(r, t)$	— mass density,
$\boldsymbol{v}(r, t)$	— vector of the velocity field
$p(r, t)$	— thermodynamic pressure,
$V_0(r, t) = \dfrac{-\gamma M_c}{r}$	— gravitational potential of the central mass M_c,
$\phi(r, t)$	— gravitational potential of circumprotostellar matter,
$\eta(r, t) \equiv \varrho\nu$	— coefficient of shear viscosity, ν is the so-called kinematic shear-viscosity,
$\eta'(r, t) \equiv \varrho\nu'$	— coefficient of compressional, viscosity, ν' is the so-called kinematic compressional viscosity,
I	— unit tensor,
$\circ$	— symbol for the dyadic product of two vectors,
$\boldsymbol{j}(r, t)$	— electric current density,
$\boldsymbol{B}(r, t)$	— magnetic induction,
$\sigma(r, t)$	— tensor of electroconductivity,
$\boldsymbol{E}(r, t)$	— electric field strength,
∇	— Nabla operator ($\overleftarrow{\nabla}$ acts as differential operator on functions to its left side),
Δ	— Laplace operator.

The above given equations will be used in different model-approximations in order to describe characteristic properties of matter and processes in circumprotostellar or circumstellar shells, envelopes and disks, as they will be discussed in the following sections.

5.2.2. The Essential Role of Magnetic Fields

As has been advocated by ALFVÉN in his pioneering planetogonic work even for a long time, magnetic fields have had an essential influence in early pre-planetary phases (for a summary of this work see ALFVÉN, 1981). There are differences in the assumed action of these early magnetic fields, but there can not be any doubt on their importance at and near to protostars. As has been mentioned in section 5.1.6., there are now strong arguments in favour of the existence of relatively strong magnetic fields of the order of $10^2 - 10^3$ G (1 G $= 10^{-4}$ Vs/m^2) in circumprotostellar shells. Consequently, also the influence of these magnetic fields (and partially their dominance) on the physical processes in the circumprotostellar environment can not be excluded. Especially the formation of the disk should have been influenced by magnetic fields (as an inner barrier).

A detailed discussion of the origin of these protostellar magnetic fields is out of the scope of this book. It shall be mentioned only, that they may originate from a magnetic flux-conservation caused amplification of interstellar fields during the collapse, as well as from dynamo effects in the circumprotostellar environment (disk dynamo), or they may result from a protostellar dynamo. The last mentioned assumption seems to be most probable, since the late phases of protostellar evolution have to be characterized by a convective energy transport in the growing protostar (cf. section 5.1.5.), but the question of origin and structure of circumprotostellar magnetic fields is yet an open task for cosmic dynamo theory.

In order to have a model for further quantitative investigations, a dipole-type magnetic field shall be assumed (for the inner regions of the (late) pre-planetary disk), having an axis, slightly tilted to the axis of rotation, but this tilt should be small. As has been shown by SHIBATA et al. (1980), a magnetosphere is formed around the stellar winds emitting protostar. The results of these investigations are shown schematically in Fig. 5.6. Region I represents the limit of a relatively weak magnetic field, where the frozen-in condition is almost satisfied. In greater distances the stellar wind is bent towards the equatorial plane. In region II flux-freezing is violated. The high temperature wind blows across the field lines. This type of solution is obtained in the limit of weak electromagnetic coupling.

In magnetospheres with a strong stellar wind, as they may be assumed for late protostars to exist (see the comments on T-Tauri stars in section 5.1.6.), region I may occupy the most part of the magnetosphere, region II may be pressed to a thin current sheet or vanish, or may become essential only at large distances from the protostar.

Region III forms a corotation zone, where the effects of gravitation and inertia are not important and the magnetic field is almost a dipole one. This steady corotation is caused mainly by the combined action of crossed electric and magnetic fields if strong electromagnetic coupling dominates in equ. (5.35). This is equivalent to $\sigma \to \infty$ in equ. (5.37), or consequently to

$$\boldsymbol{E} = -\boldsymbol{v} \times \boldsymbol{B}. \tag{5.40}$$

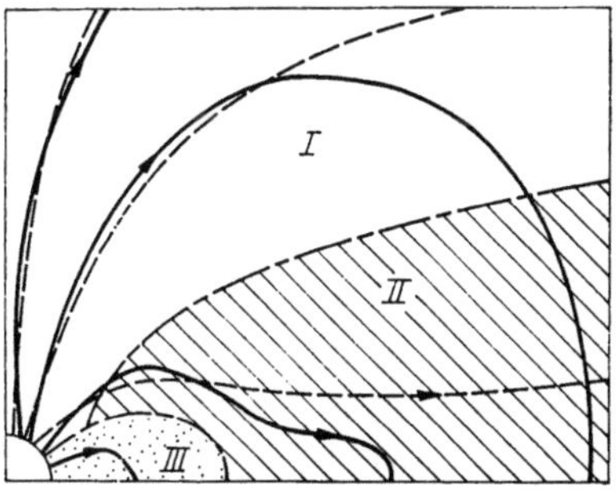

Fig. 5.6. A schematic picture of the global structure of stellar magnetospheres. The thick solid curves represent magnetic lines of force and the dashed curves the stream lines of the plasma flow. This picture is synthesized from the considerations of three extreme situations and has correspondingly three regions denoted by I, II (hatched region) and III (dotted region).

The electric field induced by the rotating dipole is then given by $\boldsymbol{E} = \boldsymbol{B} \times (\boldsymbol{\omega} \times \boldsymbol{r})$, and the resulting first order drift in the crossed electric and magnetic field is described by

$$\boldsymbol{v}_{\mathrm{D}} = \frac{\boldsymbol{E} \times \boldsymbol{B}}{B^2} = \boldsymbol{\omega} \times \boldsymbol{r}, \tag{5.41}$$

what is equivalent to corotation of the plasma with the rotating dipole (due to its unipolar induction).

Additional (drift-) velocities and currents appear, if additional forces as gravitation and inertial (centrifugal) forces are taken into account, modifying the motion of the plasma near to the late protostar. Further deviations from corotation may be caused by a decrease in the value of the electroconductivity σ, i.e. due to interactions between the plasma and the neutrals. Alfvén and Arrhenius (1976) investigated the geometry and boundaries of a region I zone by studying the equilibrium conditions under the action of the gravitational force $\boldsymbol{f}_g = -(\gamma M_c/r^2)\, \boldsymbol{e}_r$ of the central body, the rotationally caused centrifugal force $\boldsymbol{f}_c = \omega^2 r \cos \lambda \boldsymbol{e}_\varrho$ and the Lorentz force $\boldsymbol{f}_l = \boldsymbol{J} \times \boldsymbol{B}$ (see Fig. 5.7.), which is determined by $\boldsymbol{f} = \boldsymbol{f}_g + \boldsymbol{f}_c + \boldsymbol{f}_L = 0$ ($\boldsymbol{e}_\varrho$ is a unit vector perpendicular to the axis of rotation, λ corresponds to the magnetic latitude).

The components of the magnetic dipole field are given by

$$B_r = \frac{2m}{r^3} \sin \lambda, \tag{5.42}$$

$$B_\lambda = \frac{m}{r^3} \cos \lambda, \tag{5.43}$$

where m is the magnetic dipole moment. Taking into account that azimutal components of $\boldsymbol{f}_g$ and $\boldsymbol{f}_c$ do not exist, it follows

$$\frac{J_\lambda}{J_r} = \frac{B_\lambda}{B_r}, \tag{5.44}$$

indicating also the important role of field aligned currents. With $(f)_\lambda = 0$

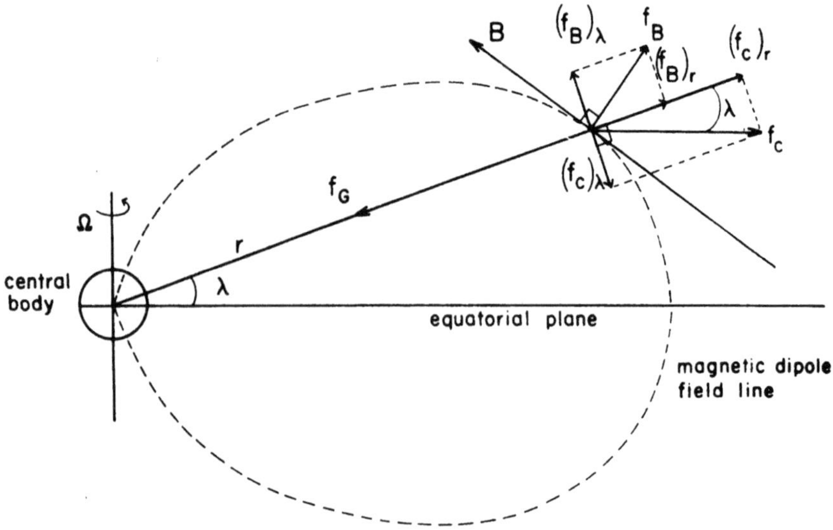

Fig. 5.7. Forces for the stationary state of partial corotation (f_g — gravitational force, f_c — centrifugal force, f_B — electromagnetic force). For the magnetic field a dipole geometry is assumed (from ALFVÉN and ARRHENIUS, 1976).

follows according to ALFVÉN and ARRHENIUS (1976)

$$f_c \sin \lambda = (f_B)_\lambda = J_\varphi \frac{2m}{r^3} \sin \lambda \tag{5.45}$$

or for $\lambda \neq 0$

$$f_c = \frac{2m}{r^3} J_\varphi . \tag{5.46}$$

Finally, from $(f)_r = 0$ follows

$$f_g = f_c \cos \lambda + J_\varphi \frac{m}{r^3} \cos \lambda , \tag{5.47}$$

and with equ. (5.44)

$$f_c \cos \lambda = 2J_\varphi \frac{m}{r^3} \cos \lambda , \tag{5.48}$$

what can not be applied to the plane $\lambda = 0$ where any rotational velocity is possible. From equ. (5.47) and (5.48) a theorem follows for the partial corotation of plasma:

The gravitational force is balanced, 2/3 by the centrifugal force and 1/3 by the electromagnetic force (ALFVÉN and ARRHENIUS, 1976).

With these results it follows for the azimutal velocity of partial corotation

$$v_\varphi{}^2 = r^2\omega^2 \cos^2 \lambda = \frac{2}{3} \frac{\gamma M_c}{r}, \tag{5.49}$$

defining with

$$r_A = \left(\frac{2\gamma M_c}{3\omega^2 \cos^2 \lambda}\right)^{1/3} \tag{5.50}$$

a surface of equilibrium, which we will call the "Alfven surface".

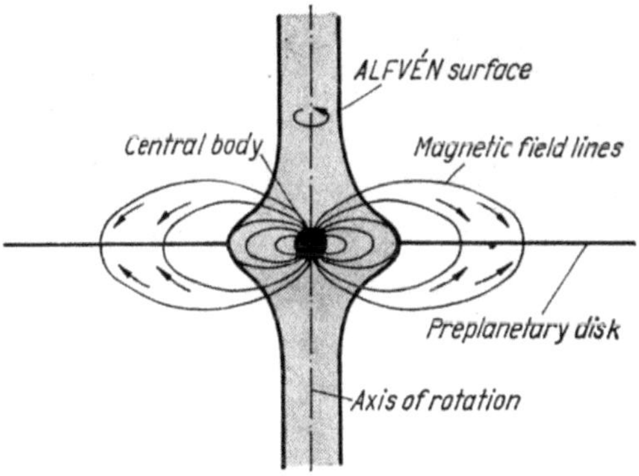

Fig. 5.8. Schematical sketch of the possible connection between biconical nebula, the state of partial corotation and the formation of the preplanetary disk.

For $r < r_A$ gravitation dominates, the plasma falls down (along the field lines) on the central body. If $r > r_A$, centrifugal force dominates, the plasma will move (along the magnetic field lines) towards the equatorial plane. Consequently, this state of partial corotation is connected with a planetogonically very essential process of the formation of a circumstellar shell at $r < r_A$ (with a radius $R = r_A$ of the shell at $\lambda = 0$). At $r > r_A$ and due to plasma drag (and with some retardation) also the neutral gas and the dust in the circumprotostellar envelope should follow the motion of the plasma towards the equatorial plane. These processes can be very effective in forming a thin late disk. They could be enhanced yet by the sedimentation process as it has been described by NAKAGAWA et al., 1981.

In earlier phases this "magnetosphere" at $r \approx r_A$ can be an effective barrier, slowing down tho inflow of matter, and leading on this way to a more massive disk.

Furthermore, it shall be noted here, that this magnetic coupling is connected with an angular momentum transport from the central body, by the stellar wind as well as by the action of the Lorentz force in envelope and disk. This results in a decrease of the angular velocity ω of rotation of the central body and a partial Joule-heating of circumprotostellar matter.

The above mentioned planetogonic conclusions seem to be in agreement with the observed fact (see section 5.1.6.), that the radius $R_{II} \doteq r_A$ of the circumprotostellar shell of T-Tauri stars increases with increasing pre-main

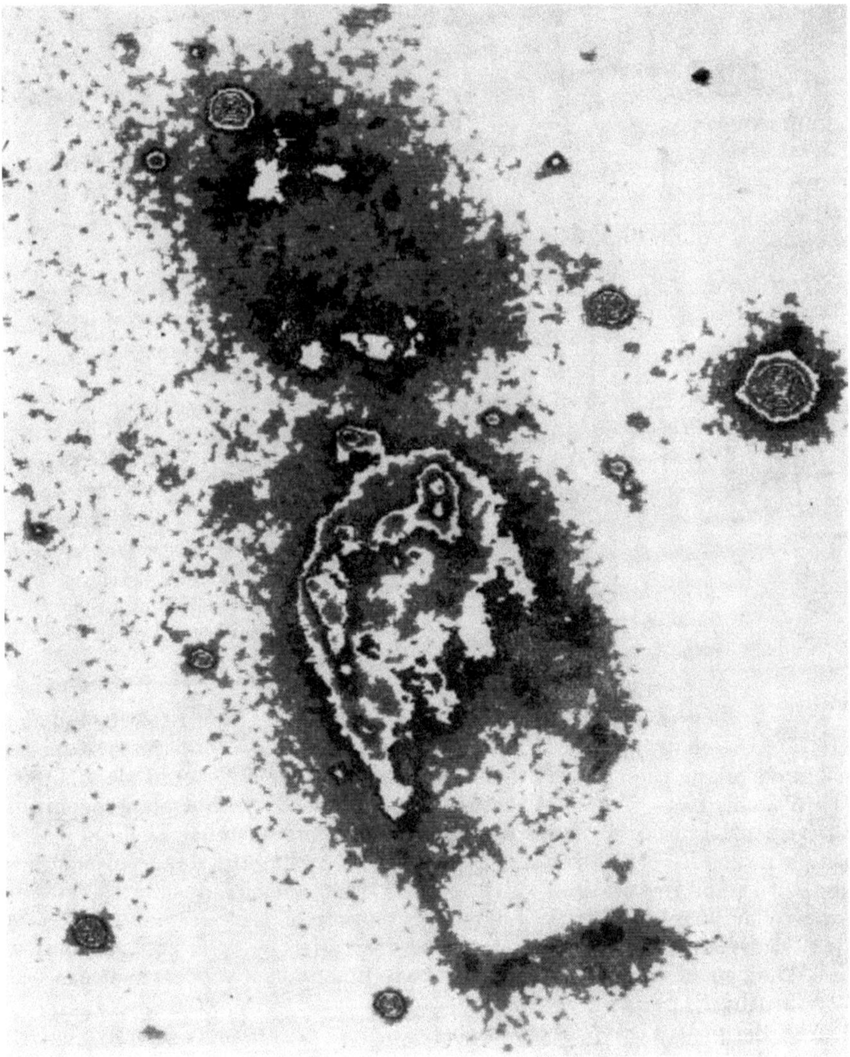

Fig. 5.3.

sequence age and that the correlated angular velocity ω (cf. equ. 5.50) decreases.

Further qualitative agreement comes from the optical appearence of bipolar nebulae, connected with young stars (see 5.1.6.), and the interpretation of these observed structures within the above given qualitative picture (see Figs. 5.8 and 5.9).

Consequently, large scale protostellar magnetic fields can play a key role in circumprotostellar environments. The influence of these magnetic fields decreases in late phases as condensation and growth of grains proceed. Then, with increasing particle mass, the action of gravitation overcomes that of magnetic fields. This can be estimated by comparing the typical frequencies of Keplerian motion Ω_K and gyration ω_g, giving

$$\frac{\Omega_K}{\omega_g} = \left(\frac{\gamma M_c}{r^3}\right)^{1/2} \frac{m}{qB} \tag{5.51}$$

with the total charge q of a particle of mass m. The electrostatic charge can be estimated by assuming a spherical particle with radius R, having the potential $U = q/4\pi\varepsilon_0 R$. Then equ. (5.51) gives with $m = \frac{4}{3}\pi R^3 \varrho_p$

$$\frac{\Omega_K}{\omega_g} = \left(\frac{\gamma M_c}{r^3}\right)^{1/2} \frac{R^2 \varrho_p}{3\varepsilon_0 BU}, \tag{5.52}$$

or for the resulting critical radius R_c for the particle

$$R_c = \left[\left(\frac{r^3}{\gamma M_c}\right)^{1/2} \frac{3\varepsilon_0 BU}{\varrho_p}\right]^{1/2}. \tag{5.53}$$

The most critical problem is to determine the electrostatic potential U of the particle, since this depends on surface properties and rotation of the particle and also on its interaction with the surrounding plasma or partially ionized gas. As an estimate a value of the order of 1 V—10 V seems to be reasonable. Then equ. (5.53) gives with a circumprotostellar magnetic induction of 1 G $= 10^{-4}$ Vsm^{-2} and a typical density of particles of $\varrho_p = 10^3$ kg m^{-3} for 1 AU conditions with $M_c = M_\odot$

$$R_c \sim 3 \cdot 10^{-6}\ \mathrm{m}, \tag{5.54}$$

indicating that particles with $R > R_c$ move in a Keplerian orbit, only slightly perturbed by plasma effects. For larger orbital radii the numerical value of R_c decreases $\sim r^{-3/4}$.

Consequently, magnetic fields should have been essential in late preplanetary phases, forming an inner plasma-dominated region (circumprotostellar shell) and influencing also the outer regions if they become sufficiently ionized. The action of these magnetic fields may be limited by the low degree of ionization far out of the circumstellar shell and when particle growth reaches dimensions of micrometers. At earlier phases, the magnetosphere can be a cause for the formation of a relatively mass-rich disk.

5.2.3. Origin of Planetary Rings

The results of the foregoing section 5.2.2. indicate that matter may accumulate for $r > r_A = (2\gamma M_c/3\omega^2 \cos^2 \lambda)^{1/3}$, i.e. outside of the Alfvén surface and near to the equatorial plane $\lambda \approx 0$. It is interesting to note, that there exists another critical distance, the so-called Roche limit $r_R = 2.44 R_0(\varrho_c/\varrho_s)^{1/3}$, inside of which the tidal action of the central body of radius R_0 and density ϱ_c destroys any satellite of density ϱ_s (Alfvén and Arrhenius, 1976). In other words, inside of r_R no accumulation to larger bodies may have occured. Consequently, "primary" planetary rings of "planetogonic" origin may have formed only in regions $r_A < r < r_R$.

Table 5.5 gives the values for r_A ($\lambda = 0$) and r_R for the Sun and the planets from Earth to Neptune (Möhlmann, 1984b).

Table 5.5. Characteristical radii in the satellite systems of the Sun, Jupiter, Saturn and Uranus

Central body	Inner satellite	r_A (10^8 m)	r_{orb} (10^8 m)	r_R (10^8 m)
Sun	Mercury	225	579	13
Earth	Moon	0.37	3.84	0.19
Mars	Phobos/Deimos	0.18	0.09/0.23	0.09
Jupiter	Amalthea	1.39	1.81	1.31
Saturn	Mimas	0.95	1.86	1.14
Uranus	Miranda	0.53	1.29	0.59
Neptune	Nereide	0.69	55.6	0.62

Here, as a typical value, $\varrho_s = 1.5 \cdot 10^3$ kg m^{-3} has been used for the satellites of Saturn, Uranus and Neptune, while $\varrho_s = 3 \cdot 10^3$ kg m^{-3} was choosen for the satellites of the inner planets.

The essential result is that only for Saturn and Uranus the above mentioned condition for r can be satisfied, since only in these two cases the condition $r_R > r_A$ is fulfilled. This may be a simple explanation for the fact that only these two planets have "original" rings. A further conclusion (prediction) is that Neptune has no "primary" rings because of $r_R < r_A$ for Neptune.

5.2.4. Neutral Gas Processes

Due to the low degree of ionization of interstellar matter, neutral gas plays an important role in collapse processes (which in their latest phases may be halted at $r = r_A$) and also for the evolution of the circumprotostellar environment, if a cogenetic origin of circumprotostellar matter and the protostar is assumed (as it is done throughout this book). The originally dominating neutral gas may evolve in these late phases on different ways, as

- falling down on the Alfvén surface and then on the central body,
- moving outward due to the action of angular momentum transport and a strong stellar wind or radiation pressure,
- condensation and deposition at grains (as nucleation centers),
- ionization, correlated with a further outward motion (due to the above mentioned processes) or condensation from the plasma phase. The last mentioned possibility indicates the essential but yet open role of plasma-chemistry for a better understanding of early preplanetary processes.

Obviously, and particularly in earlier phases and in outer regions of the circumprotostellar envelope ($r > r_A$), physical processes connected with the neutral gas were most important. Therefore, it is the aim of this section to investigate the importance of processes which are influenced by neutral gas.

One important consequence of the presence of a gas is its thermal pressure. In order to study its influence, magnetic fields, friction and self-gravitation of circumstellar matter shall be neglected for the moment. Then it follows from equs. (5.34) and (5.35) for steady conditions

$$(\boldsymbol{v}\nabla) \circ \boldsymbol{v} = -\boldsymbol{e}_r \frac{\gamma M_c}{r^2} - \frac{1}{\varrho} \operatorname{grad} p, \tag{5.55}$$

giving for axisymmetric conditions $\boldsymbol{v} = v_0(r)\, \boldsymbol{e}_\varphi$

$$\frac{v_0^2}{r} = \frac{\gamma M_c}{r^2} + \frac{1}{\varrho} \frac{\partial}{\partial r} p \tag{5.56}$$

or

$$v_0 = \left\{\gamma \frac{M_c}{r} + \frac{r}{\varrho} \frac{\partial}{\partial r} p\right\}^{1/2} = \Omega_{KP} r \tag{5.57}$$

with

$$\Omega_{KP} = \left\{\frac{\gamma M_c}{r^3} + \frac{1}{\varrho r} \frac{\partial}{\partial r} p\right\}^{1/2}. \tag{5.58}$$

Consequently, pressure is able to modify the large-scale dynamics of circumstellar matter. The characteristic length-scale L_p for pressure gradients to be effective can be estimated from equ. (5.58) with $p = \varrho \dfrac{k}{m_H} \cdot T$ by

$$L_p \sim \frac{kr^2}{\gamma m_H M_c} \cdot T \tag{5.59}$$

with $m_H \approx 10^{-27}$ kg for a hydrogen cloud with a temperature of about $T \sim 10^3$ K follows for the present Solar system $L[m]/r[m] \approx 10^{-13} r[m]$, indicating that at 1 AU-distances pressure variations with scales equal or smaller than 10^9 m may overcome the gravitational action of the central body. At scales of the order of 10^{10} m the effects of pressure and gravitation of the central mass become comparable. This can be understood as an indication for the outer boundary of circumprotstellar matter (belonging to the system) to be at distances of the order of 10^{13} m. This scale can have been larger at earlier phases, when the mass of the yet growing Sun was smaller. This could be of relevance for the origin of comets.

At the smaller scales (wavelengths) acoustic and gravo-acoustic waves may have been evolved, which became unstable if there was an appropriated radial density profile $\varrho(r)$. This should have been the cause for nonlinear processes to develope. In other words turbulence may have been evolved, what might have been of importance also for the larger scales and also for the growth if grains (Völk, 1981, 1984). Eventually, the level of turbulence has been (regionaly) reduced by the action of magnetic fields as well as by interactions with more massive dust-particles. Consequently, at this early (gas-dominated) state of preplanetary evolution, pressure can have been very essential, counteracting any structuring processes in a low-density disk. On the other side and as will be shown later, local gravitation overcomes the dispersive action of pressure, if the density of disk matter is of the order of or greater than $\varrho \approx c^2/4\pi\gamma R^2$ (when R is the scale length for a disturbance in the density). Consequently, gravitational instability and the growth of protoplanets can have been possible then.

A further essential influence on the dynamics of the preplanetary disk of gas and dust particles comes from the dissipative action of frictional forces, caused by turbulence and by collisions of molecules and particles, moving (on the average) on Keplerian orbits (it is shown in the next section that the dust follows the azimutal motion of the gas). A yet unresolved question is a precise determination of the corresponding "kinematic viscosity" $\nu = \eta/\varrho$. Following Lynden-Bell and Pringle (1974), $\nu = Re^{-1} r v_\varphi$ can be assumed, what results from the assumption, that the pure smooth Keplerian flow is likely to be disturbed by turbulent eddies, and that these eddies are fed with energy until the eddy viscosity reaches that value at which the effective Reynolds number is reduced to the critical one for the onset of turbulence. Thus $\Omega_K r^2/\nu_{\text{eddy}} \approx Re$. From experience in other fluid dynamical problems Lynden-Bell and Pringle (1974) conclude that it is reasonable to take $Re \approx 10^3$. Therefore the value $\nu = 10^{-3}(r\gamma M_c)^{1/2}$ will be used for the estimations made, throughout this book, but a more precise derivation is yet a task for future (for alternative approaches see Cassen and Moosman, 1981). Having in mind these uncertainties in the value of ν it is not surprising that the importance of large-scale frictional processes for the dynamics of the protostellar environment is in general a yet open problem.

The above given value for the kinematic viscosity corresponds to a dissipation time $\tau_0 \approx r^2/\nu \approx 10^3\,\Omega_K{}^{-1}$, being about 10^3 times greater than the Keplerian period. In principle the same result follows of $\nu \approx 0.1$ cl is taken as it follows from Morfill et al., 1985. Therefore, as a first approximation it can be assumed for preplanetary conditions that the (possibly pressure-infuenced) Keplerian motion dominates and that friction can be treated as a disturbance, having a relatively large time-scale. Especially it has to be noted that the timescale of radial and azimutal structuring gravitational processes has had to be smaller than that of the above discussed dissipative processes, when these structuring processes were effective.

The essential properties of frictionally caused dissipative processes in the preplanetary disk shall be discussed now more in detail. In the approximation that pressure can be neglected equ. (5.35) can be reduced to

$$\boldsymbol{k} = -\mathrm{grad}\ V_0 + \mathrm{div}\ \mathbb{P}, \tag{5.60}$$

where with $v_{0\varphi} = \boldsymbol{e}_\varphi(\gamma M_c/r)^{1/2}$ follows

$$\mathrm{div}\ \mathbb{P} = -\boldsymbol{e}_\varphi \frac{3}{2} \frac{(\gamma M_c)^{1/2}}{r^2} \frac{\partial}{\partial r} \varrho(r, t)\, \nu(r, t)\, r^{1/2}. \tag{5.61}$$

The balance equation (5.38) for angular momentum density yields than (cf. Hayashi, 1981)

$$v_r = -\frac{3}{\varrho r^{1/2}} \frac{\partial}{\partial r} \nu \varrho r^{1/2} \left(= -\frac{3\nu}{2r} \text{ for } \eta = \text{const.}\right) \tag{5.62}$$

and the equation of continuity becomes

$$\frac{\partial}{\partial t} \varrho = \frac{3}{r} \frac{\partial}{\partial r} r^{1/2} \frac{\partial}{\partial r} \nu \varrho r^{1/2}. \tag{5.63}$$

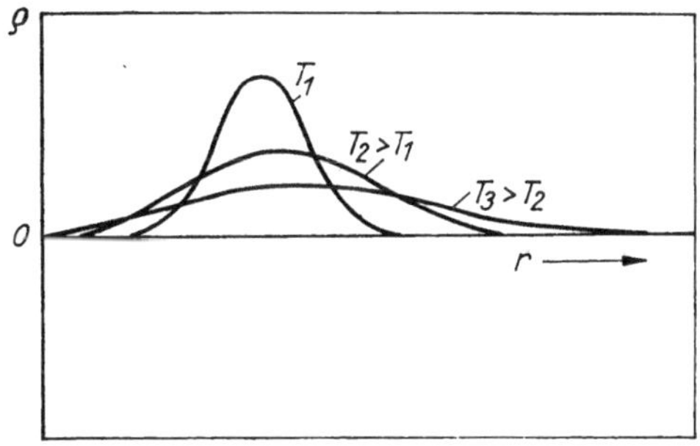

Fig. 5.10. Schematical picture of the dissipative decay of a ring due to the action of viscosity (T = time).

This is a diffusion type equation, indicating together with equ. (5.62) that an outward diffusion of matter occurs for a sufficiently strong radial decrease in density, while inward diffusion follows for a radially increasing $\nu\varrho r^{1/2}$. This is illustrated in Fig. 5.10. More detailed investigations of the diffusion-like action of mechanical friction have been made by Lüst (1952) and Lynden-Bell and Pringle (1974). They have shown, that friction causes mainly an inward motion of matter (as well as an outward flow, which is necessary to conserve total angular momentum).

In this connection it should be mentioned, that a negative viscosity would lead to a "negative diffusion", what would be equivalent to an increase of originally weak structures. This very interesting process has been discussed in the literature to understand the formation of preplanetary jets from earlier bands or rings (Alfvén, Arrhenius, 1976) and of fine-structures in the Saturnian rings (Steward et al., 1984). But it should be noted, that the working of this process of negative viscosity is connected with some strong constraints to the mechanical properties of the colliding particles. The physical content of these restrictions is not yet clear today.

In the above mentioned detailed mathematical analysis, Lynden-Bell and Pringle (1974) found for an initial distribution of

$$2\pi R^3 \nu \sigma \left. \frac{d\Omega_K}{dr} \right|_{t=0} = C(\gamma M_c r)^{1/2} \, e^{-a(\gamma M_c r)^2} \tag{5.64}$$

and with the abbreviation

$$T' = 12(\gamma M_c)^2 \nu a t + 1 \tag{5.65}$$

that the radial velocity is given with

$$v_r = -\frac{3\nu}{2r} \left\{ 1 - \frac{4a(\gamma M_c r)^2}{T'} \right\} \tag{5.66}$$

and the surface mass-density can be described by

$$\sigma = C \frac{(T')^{-5/4}}{3\pi\nu} \exp\left\{ -\frac{a(\gamma M_c r)^2}{T'} \right\}. \tag{5.67}$$

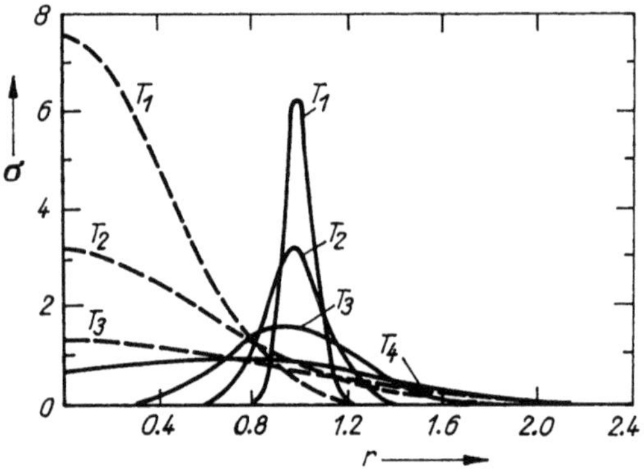

Fig. 5.11. Viscosity caused decay of structures in a thin preplanetary disk for different times $T_3 > T_2 > T_1$ (from Lynden-Bell and Pringle, 1974).

In addition to the diffusion-type character of the solution it becomes clear by these more detailed results, that the radial flux ϱv_r (and v_r) is directed outwards for $r_K > (T'/4a)^{1/2}/\gamma M_c$ and inwards inside this time-dependent radius r_K. This point of velocity reversal moves outwards with increasing time and overtakes more and more material. These results are illustrated in Fig. 5.11 together with those for a δ-function initial mass distribution with initial mass M_1 of the disk (ring) at $r = r_1$. For the case of an inner corotating magnetosphere, which is sufficiently strong to resist the influx of matter at $r = r_A$, Lynden-Bell and Pringle (1974) found the analytical solutions

$$v_r = -\frac{3\nu}{2r} \left\{ \left[1 - \left(\frac{r_A}{r} \right)^{1/2} \right]^{-1} - 2 \frac{r^2}{r_1^2 T'} \right\} \quad (r_A \ll r_1) \tag{5.68}$$

and

$$\sigma = \frac{M_1(T')^{-5/4}}{2\pi r_1{}^2} \left[1 - \left(\frac{r_A}{r}\right)^{1/2}\right] \exp\left(-\frac{1}{2T'}\frac{r}{r_1{}^2}\right). \tag{5.69}$$

A remarkable conclusion that can be taken from these results is, that in the case of formation of a sufficiently strong circumprotostellar magnetosphere (in late protostellar phases), the circumprotostellar matter accumulates mainly close to (but outside) the boundary $r = r_A$, with a rapid radial decrease in density. In order to conserve angular momentum a small amount of matter survives at distance $r \gg r_A$. The further evolution of this matter near to $r = r_A$ depends on the possible action of instabilities (of RAYLEIGH-TAYLOR type) or field aligned flows along the magnetic field lines, which could be able to cause a continuation of an inward flow onto the central body. This is a yet open problem.

Otherwise, if there is no resistance for the inward flow, nearly all the matter moves onto the central body as long as viscosity is effective.

Structuring processes are restricted as long a viscous interactions may influence the large-scale dynamics of preplanetary matter. The onliest exception could be the above mentioned special mode of "negative diffusion", as it has been advocated by ALFVÉN and ARRHENIUS (1976). But it should be noted, that structuring processes can be effective if their time-scale is much shorter than that of viscous dissipation and if the evolved structures (as "protoplanets") are not influenced by viscous decay.

Very probabely, a further consequence of the existence of turbulence and viscosity in early preplanetary matter are enhanced transport processes leading at least to a mass loss (decrease of density) of the earlier more massive disk. But to be complete, it must be noted as an additional possibility for structuring processes (connected with viscosity) that a coupling of resonance effects and viscous interactions should be able too to generate and maintain structures. The importance of such resonances has been investigated especially by TORBETT et al. (1982), WARD (1981) and WEIDENSCHILLING and DAVIS (1985). The main idea is that resonant interactions with a perturbing massive planet act opposite to the orbital decay (due to gasdrag) and may therefore lead to a trapping in stable orbits. This can indeed have been in late phases after the formation of protoplanets. These investigations shall not be discussed here in detail. This is because the main aim of this book is to study the possible influence of gravitational instabilities of disks and rings in forming protoplanets.

Concluding this section it shall be stated again, that the evolution of circumprotostellar matter is essentially influenced by viscosity (cf. the actual work of TSCHARNUTER (1985) and MORFILL (1985)). After first formation phases, as they were described in chapter 5.1., vertical infall and viscosity determine the time scale for further growth of the central body and of the further evolution of the forming circumprotostellar disk. The time scale τ_f for viscous processes can be estimated with $\tau_f \approx L^2/\nu \approx R^2/\nu$. For gravitationally structuring processes to become effective, their time scale $\tau_g = 2\pi/\omega_g$ (with $\omega_g{}^2 = 4\pi\gamma\varrho$, see section 5.3.2.) has to be shorter than τ_g. Conditions, favouring the formation of such relatively mass-rich disks are a sufficient large amount of original angular momentum of the prestellar cloud, a relatively high degree of original flatness of this cloud and an externally caused preenhancement of the density of this cloud. Furthermore, the above mentioned "inner barrier" can be an additional cause by damming up the inward mass flow.

A further consequence of effective viscosity and high mass density (with a resulting optical thick disk) is that the disk is heated by friction due to radially differential rotation (Keplerian motion) and cooled only at it's surfaces. This leads with an height-integrated mass density $\sigma = \int dz\varrho$ to a surface temperature T_s

$$T_s^4 = 9\nu\sigma\Omega_K^2/8\sigma_{SB}, \tag{5.70a}$$

where $\sigma_{SB} = 5.669 \cdot 10^{-8}\ W/m^2\ deg^4$ and $\Omega_K^2 = \gamma M_c/r^3$. Consequently, the surface temperature ranged between $20K$ in the outer planetary and $2000K$ in inner planetary regions. Inside the opaque disk the temperature should be slightly greater, what can be a cause for convectively driven turbulence.

A more detailed numerical model for the radial dependence of disk-temperature has been given by LIN and PAPALOIZON (1986). Quantitatively, their result can be approximated with $T[K] = 10^{+14}r^{-1}[m]$, leading to $T \approx 1670\ K$ at the pre-Merkurian region and to $T \approx 20K$ at pre-Neptunian distance. This "thermal model" shall be used in the more quantitative sections of this book. With $\bar{\gamma} = 1.40$ and for an H_2-gas this model gives

$$c^2 = \bar{\gamma}\,\frac{k}{m_{H_2}}\,T = 5.77 \cdot 10^{17} r^{-1}[m]. \tag{5.70b}$$

This is equivalent to $c^2[m^2/s^2] = c_0^2/r[m]$ with $c_0^2 = 5.77 \cdot 10^{17}$.

Summarizing, the neutral gas dominated preplanetary phases can be characterized by

- the dominance of early inward flows onto the growing protosun from a massive disk,
- an outer boundary at $L \sim 10^{13}$ m for the Solar system,
- an frictionally caused internal transport and mass loss by inward fluxes onto the central body (if viscosity is sufficiently great),
- an outward directed mass flow towards regions where viscosity is not effective and due to vertical outflow,
- a slowing down of the inward flow of matter due to the existence of an inner "barrier" (magnetosphere of hot plasma),
- a radial temperature profile, which can be approximated well with $T \sim r^{-1}$,
- a turbulence-caused growth of grains up to larger blocks (VÖLK, 1981, 1984),
- the possibility for fast gravitationally structuring processes with final products, which are not essentially influenced by friction. A necessary condition is that these structuring processes, were faster than the dispersive action of friction. This implies a sufficiently high mass density of the circumprotostellar matter.

5.2.5. Dust-Gas Interactions

An essential process, connected with the existence of gas in circumprotostellar environments is the frictional interaction between the gas (with thermal velocity $v_{\text{th}} = \left(\frac{3kT}{m}\right)^{1/2}$) and (solid) particles (for a summary of these processes see WEIDENSCHILLING, 1977 and WILLIAM and CRAMPIN, 1971) what is caused mainly by the reduced angular velocity of the gas due to the action of pressure (see equ. 5.58). The resulting velocity difference (for realistic small pressure gradients, if compared with the action of gravity) amounts to

$$\Delta v = \frac{r}{2\varrho_0} \frac{\partial p_0}{\partial r} \left(\frac{r}{\gamma M_c}\right)^{1/2}, \tag{5.71}$$

when simple Taylor-expansion is used.

The characteristic time t_e in which Δv would be reduced by a factor e by a decelerating constant drag force according to Epsteins drag law (WEIDENSCHILLING, 1977a)

$$f_d \approx \frac{4}{3} \pi \varrho R^2 v_{\text{th}} \Delta v \tag{5.72}$$

is given then via

$$\frac{4}{3} R^3 \pi \varrho_p \, \text{d} \frac{\Delta v}{\Delta t} = f_d \tag{5.73}$$

by

$$t_e = \frac{\varrho_p R}{\varrho v_{\text{th}}}, \tag{5.74}$$

where ϱ is the mass density of the gas and ϱ_p that of the individual particle. Epstein's drag law is valid for the case of $|\Delta v| \ll v_{\text{th}}$ and $R \ll \lambda$, where λ is the free path of the gas molecules. With typical values $v_{\text{th}} \sim 10^3 \text{ ms}^{-1}$, $\varrho \sim 10^{-6} \text{kg m}^{-3}$, $\lambda \approx 10^{-1}$ m and $\varrho_p \sim 10^3 \text{ kg m}^{-3}$ it follows from (5.74) that small particles very quickly have to follow the (slightly slower) azimutal motion of the gas. Consequently they tend to fall into the direction of the central body (radial sedimentation). The maximum velocity is reached for

$$\frac{4}{3} \pi \varrho R^2 v_{\text{th}} (\Delta v)_r \approx \gamma \frac{M_c \frac{4}{3} \pi R^3 \varrho}{r^2} \tag{5.75}$$

giving

$$(\Delta v)_r \approx \frac{\gamma M_c R}{r^2 v_{\text{th}}}. \tag{5.76}$$

Integration with time gives for the characteristic time t^* of a particle of radius R for falling from an initial radius R^* (size dependent radial sedimentation)

$$t^* = \frac{v_{\text{th}}}{3(\Omega_K{}^*)^2 R} \tag{5.77}$$

where $\Omega_K{}^* = [\gamma M_c/(R^*)^3]^{1/2}$. With $R^* \approx 10^{11}$ m and $T \approx (10^2-10^3)$ K follows

$$t^*(s) \approx 10^{15} R^{-1}(\text{m}). \tag{5.78}$$

This indicates that particles with diameters smaller than some meters can not have been sedimented inwards essentially by this process in time scales of the orders of (10^5-10^6) years.

The above given estimation can be applied only if the mean free path of the molecules of the gas exceeds the dimensions of the dust particles or small bodies respectively. The mean free path l can be estimated via $l = 1/N\sigma$ (N — particle number density, σ — effective cross section) to be of the order of 10^{-4} m for $\varrho \sim 10^{-3}$ kg/m^3 and 10^2 m for $\varrho \approx 10^{-9}$ kg/m^3. Consequently, for m-sized bodies Stokes law has to be used to describe the frictional interaction between a body on a Keplerian orbit and the weakly slowlier revolving gas. With Stokes low follows

$$m_b \frac{d}{dt} \Delta v = -6\pi\eta R \,\Delta v \tag{5.79}$$

where $m_b = \frac{4}{3} \pi \varrho_p R^3$ and the coefficient of friction η.

This coefficient η is given by $\eta \approx (\varrho_{gas})\, v_{th} l/3$. This is of the order of 10^{-4} kg/ms for conditions in the preplanetary disk. Consequently, the decay time t_e is in this case given by

$$t_e[s] = \frac{m_b}{6\pi\eta R} = \frac{2\varrho_p R^2}{9\eta} \approx 2 \cdot 10^{+6} R^2\,[\text{m}]. \tag{5.80}$$

Essential changes in the ortibal radius can be expected therefore for bodies with 1-m-radius very soon, km-sized bodies have a decay-time of the order of 10^5 years while that of 100-km-diameter bodies exceeds that of the age of the Solar system. An interesting condition is that planetary sized bodies were not essentially influenced in their orbits by the gas.

Another result of gas-dust interactions is a vertical dust sedimentation in the equatorial plane of a disk. The equation of vertical motion v_z for a dust particle under the influence of the vertical component of gravitation of the central body and the Epstein drag (cf. equ. 5.72) is given by

$$m_p \frac{dv_z}{dt} = -\frac{4}{3} \pi R^2 \varrho v_{th} v_z - \gamma \frac{m_p M_c}{r^3} z, \tag{5.81}$$

where z is the vertical coordinate. The equilibrium is determined then by

$$v_z = -\frac{R}{v_{th}} \frac{\varrho_p}{\varrho} \Omega_K{}^2 z. \tag{5.82}$$

The time scale for vertical sedimentation $\tau_s \approx |H/v_z|$ from $z = H$ is given then with

$$\tau_s = \frac{\varrho}{\varrho_p} \frac{v_{th}}{R} \Omega_K{}^{-2}, \tag{5.83}$$

what is of order the of 10^3 years for $R = 10^{-3}$ m particles and of 10^6 years for $R = 10^{-6}$ m particles, indicating that this process of vertical sedimentation of dust may be relatively fast and effective. This sedimentation-generated thin dusty disk was appearently an essential structure of the circumprotostellar disk.

A result of possible dust-dust interactions consists in a coagulative growing of the dust particles in course of their vertical sedimentation (see section 5.2.4., and due to turbulence (see VÖLK, 1981, 1984)). Following the line of SAFRONOV (1969) and NAKAGAWA et al. (1981), the mass-increment of a dust grain of Radius R due to coalescence in a vertical displacement by sticking (with the probability p_s) is given by

$$dm = p_s \pi R^2 \varrho_d \, dz \tag{5.84}$$

where ϱ_d is the mean mass density of the dust. In the same way it follows with $dm = 4\pi R^2 \varrho_p \, dR$

$$\Delta R = \frac{\varrho_d p_s}{4\varrho_p} \Delta z . \tag{5.85}$$

Numerically this gives for 1 AU conditions with

$$\varrho_d = 0.0034 \varrho_g \approx 2 \cdot 10^{-8} \text{ kg m}^{-3} \tag{5.86}$$

(PODOLACK and CAMERON, 1974; KUSAKA et al. 1970),

$$\varrho_p = 3 \cdot 10^3 \text{ kg m}^{-3} \tag{5.87}$$

and

$$z = 6 \cdot 10^9 \text{ m} \tag{5.88}$$

(KUSAKA et al., 1970) the value

$$\Delta R = p_s \cdot 10^{-2} \text{ m} \tag{5.89}$$

with ($p_s \sim 1$).

Consequently, at 1 AU the sedimenting particles have radii of the order of centimeters when they are settled in the equatorial plane. This value may increase, if the relative dust contribution is greater than the assumed value of $\varrho_d/\varrho_p \sim 0.0034$.

The growth process of dust grains up to bodies of greater diameters has been investigated by VÖLK (1981), MORFILL and VÖLK (1984), MORFILL et al. (1985). Their quantitative treatment of this problem has been based on the model of a steady but turbulent protostellar disk feeding yet the growing protosun. Turbulence in the relatively thin gas causes the constituents to diffuse spatially. If there exists a large-scale temperature gradient across the turbulent region, this diffusion will lead to partial evaporation of dust grains deep inside the disk and recondensation of the resulting vapour further out. Furthermore, this diffusion should result in a diffusive mixing of thermally altered grains with incoming fresh grains and particles. The turbulent gas motions cause relative grain motions (coagulation, eventually fragmentation). The diffusion of larger grains is modified by the above described inward drift relative to the smaller grains and to the grain vapour. A yet open problem is the applicability of these investigations to disks of higher density.

Summarizing this section it can be stated that, as long as gas is important for the dynamics of dust and particles

- the dust follows (with some retardation) the azimutal motion of the gas,
- a size-dependent radial sedimentation takes place. This is connected with a resulting chemical fractionation,
- dust particles with dimensions up to some ten centimeters or meters fell down onto the central body while planetary sized bodies were able to survive at their orbits. This leads to a radial gradient in the chemical composition, leaving volatiles in the outer regions and accumulating heavier elements in the inner region of the disk,
- vertical sedimentation and particle growth
- turbulence-caused growth acts into the same direction with yet greater final products (VÖLK, 1981).

5.2.6. Radiation Pressure

An essential effect, influencing the dynamics of the neutral gas (and the possibly connected plasma) and small particles of the dust is, that in late phases the pressure of the electromagnetic radiation from the protostar becomes essential if the gradually thinning gas becomes optically thin. Consequently, this discussion refers mainly to late planetogonic phases. If the energy flux is given by S, it follows for nonrelativistic radial velocities of the particles of radius R_p and mass $m = \frac{4}{3} \pi R_p{}^3 \varrho_p$ for the radial force of the radiation pressure and the radial distance r

$$f_{rad} = \frac{\pi R_p{}^2 S(r)}{c}. \tag{5.90}$$

This is equal to the gravitational force for

$$R_p = \frac{3r^2 S}{4\gamma M_c \varrho_p c}. \tag{5.91}$$

At 1 AU ($r = r_0$) and with $S_0 = 1.395 \cdot 10^3$ W m^{-2} (solar constant) this gives numerically

$$R_p \sim 6 \cdot 10^{-7}\ \text{m}, \tag{5.92}$$

what is independent of r due to $Sr^2 = \text{const.} = S_0 r_0{}^2$.

Here, typical values of $\varrho_p = 10^3$ kg/m^3 and $M_c = M_\odot$ are used. Consequently, particles with radii smaller than $6 \cdot 10^{-7}$ m (submicron particles) will be repelled by the radiation of the protostar. This effect can be an essential cause for the gas and sub-micron grains to be moved out of the system or into its outer regions. This transport can be supported by a strong and high temperature stellar wind from the young star, which, similar to the electromagnetic radiation, is able to ionize the gas (by Alfvén's critical velocity phenomenon) as well as to accelerate dust and gas outward.

Furthermore, it shall be noted here, that the remaining particles are heated due to the protostellar and stellar radiation, getting an equilibrium temperature T_p, that according to $T_p^4 \sim r^{-2}$ has a radial dependence $T_p \sim r^{-1/2}$. Similar to comets, this temperature is sufficient for volatile gases to evaporate inside of $r \approx 10^{12}$ m. This could have led to a prolongation of the time of gas flow (of volatiles) from the inner parts of the system to its outer regions and might be a cause for the different chemical composition of smaller bodies in the inner and the outer parts (asteroids, comets).

It is interesting to note here, that there is an additional effect of electromagnetic radiation, influencing the motion of particles. This is the so-called Poynting-Robertson effect. This takes into account the azimutal deceleration of an azimutally moving body due to radiation pressure, what can be described by

$$f_{az} = \frac{v_\varphi}{c} \pi R^2 S/c, \tag{5.93}$$

giving

$$\frac{\mathrm{d}mv_\varphi}{\mathrm{d}t} = (mv_\varphi) \frac{3S}{4RS_p c^2}. \tag{5.94}$$

Hence, during a time τ_e the orbital momentum decreases by a factor of e. Numerically it follows for 1 AU conditions with $\varrho_p = 10^3$ kg m^{-3} that

$$\tau_e = R\,\frac{4\varrho_p c^2}{3S} \approx 10^{17} R\ (\mathrm{m}) \tag{5.95}$$

is the characteristic time scale for falling down to the Sun. It is interesting to note, that it follows from $\tau_e \approx 5 \cdot 10^9$ years $= 1.5 \cdot 10^{17}$ s, that "original" particles with $R < 1$ m did not survive from the early phases of the planetary system up to the present time due to the Poynting-Robertson effect.

5.2.7. Gravitational Disk Potential

The gravitational potential $\phi_0(\boldsymbol{r})$ of a given mass distribution $\varrho(\boldsymbol{r})$ can be found in principle by using Poisson's equation

$$\Delta\phi_0 = 4\pi\gamma\varrho \tag{5.96}$$

or from its sometimes more appropriated integral representation

$$\phi_0(\boldsymbol{r}) = -\gamma \int\limits_V \mathrm{d}\tau' \frac{\varrho(\boldsymbol{r}')}{|\boldsymbol{r} - \boldsymbol{r}'|}, \tag{5.97}$$

what in the case of a surface density $\sigma(\boldsymbol{r})$ reduces to

$$\phi_0(\boldsymbol{r}) = -\gamma \int\limits_{S'} \mathrm{d}f' \frac{\sigma(\boldsymbol{r}')}{|\boldsymbol{r} - \boldsymbol{r}'|}, \tag{5.98}$$

giving in cylindrical coordinates for the potential along the z-axis of a constant and axisymmetric surface density σ_0 in the plane $z = 0$ and inside a radius "a"

$$\phi_0(z) = -\gamma\sigma_0 \int_0^{2\pi} d\phi \int_0^a \frac{r\,dr}{(r^2 + z^2)^{1/2}}, \tag{5.99}$$

and consequently

$$\phi_0(z) = -2\pi\gamma\sigma_0\{(a^2 + z^2)^{1/2} - |z|\}. \tag{5.100}$$

This representation can be used to identify the coefficients A_l and B_l of Legendre polynomials (in spherical coordinates) in the (external) solution

$$\phi_0(r,\delta) = \sum_{e=0}^{\infty} \left(A_l r^l + \frac{B_l}{r^{l+1}}\right) P_l(\cos\vartheta) \tag{5.101}$$

of equ. (5.98) for axisymmetric mass distributions. In order to compare equ. (5.101) for $\vartheta = 0$ (i.e. for the z-axis) and equ. (5.100), the last mentioned one has to be represented by Taylor expansion in terms of powers of z/a and a/z for $z < a$ and $z > a$ respectively, giving then for $r < a$ in spherical coordinates

$$\phi_0(r,\vartheta) = -2\pi\gamma\sigma_0 a \left\{1 - \frac{r}{a}|P_1| + \frac{1}{2}\left(\frac{r}{a}\right)^2 P_2 - \frac{1}{8}\left(\frac{r}{a}\right)^4 P_4 \right.$$
$$\left. + \frac{1}{16}\left(\frac{r}{a}\right)^6 P_6 - \frac{5}{128}\left(\frac{r}{a}\right)^8 P_8 \pm \cdots\right\}, \tag{5.102}$$

while for $r > a$ follows

$$\phi_0(r,\vartheta) = -2\pi\gamma\sigma_0 a \left\{\frac{1}{2}\frac{a}{r} - \frac{1}{8}\left(\frac{a}{r}\right)^2 P_2 + \frac{1}{16}\left(\frac{a}{r}\right)^5 P_4 - \frac{5}{128}\left(\frac{a}{r}\right)^7 P_6 \pm \cdots\right\}. \tag{5.103}$$

Comparing now the potential Φ_0 of a disk with that of a central body with mass M_c, it follows, that to describe the ground state, an influence of the disk potential in the radial directions (at $z \approx 0$) can be neglected in a first approximation as long as

$$\sigma_0 < \frac{M_c}{\pi \cdot a^2}. \tag{5.104}$$

This condition is equivalent to the restriction on disks with $m_d < M_c$.

5.2.8. Ground States of Disks

For a gaseous disk of a massive central body (of mass M_c), equ. (5.34) reduces to

$$\frac{\partial \boldsymbol{v}}{\partial t} + (\boldsymbol{v}\,\text{grad})\,\boldsymbol{v} = -\text{grad}\,(V + \Phi) - \frac{1}{\varrho}\,\text{grad}\,p, \tag{5.105}$$

which together with Poisson's equation, $V = -\gamma \frac{M_c}{r}$, and the equations of state and of continuity completes a closed set of equations. A stationary and axisymmetric "ground state" (cylindrical coordinates are used with the disk-plane $z = 0$) is given then with $v_z = 0$ and $\boldsymbol{v} = e_\varphi v_0(r)$ at $z = 0$ by

$$v_0(r) = \pm \left\{\gamma \frac{M_c}{r} + r \frac{\partial}{\partial r} \Phi_0 + \frac{r}{\varrho_0} \frac{\partial}{\partial r} p_0\right\}^{1/2}, \tag{5.106}$$

where Φ, ϱ_0 and p_0 indicate the "ground state".

This reduces for low mass disks and for pressure gradients (of ground states), satisfying the relation $L_p/r > 10^{-13} r[m]$ (see equation (5.59)), to

$$v_0 = \sqrt{\gamma \frac{M_c}{r}}. \tag{5.107}$$

This result will be used in this book. The modification due to Φ_0 will not be essential at $z = 0$ for a smooth radial profile of the disk density also for $m_d \lesssim M_c$.

Two limiting cases for disk-like ground states shall be discussed now more in detail, namely first that of a (vertically) "gravitationaly bound" disk, where the gravitation of a massive thin disk determines its vertical structure and then that of a "Keplerian disk", where the vertical structure is determined by the gravitation of the central body. For the "gravitationaly bound disk" follows for $z \neq 0$ for the (isothermal) static ground state

$$0 = -\varrho_0 \frac{\partial \Phi_0}{\partial z} - \frac{\partial}{\partial z} p_0, \quad \frac{\partial^2}{\partial z^2} \phi_0 = 4\pi\gamma\sigma_0 \tag{5.108}$$

if the vertical scales are much smaller than the horizontal scales. Equation (5.108) is valid as long as the z-component of the gravitation of the central body can be neglected, compared with that of the disk. From (5.108) follows for the vertical density profile of the disk if c^2 is the isothermal sound speed that

$$\varrho_0 = \frac{K}{\cos h^2(z/\sqrt{2}\, H)} \tag{5.109a}$$

where

$$H^2 = c^2/4\pi\gamma K = \frac{c^2}{\omega_g^2(z = 0)}. \tag{5.109b}$$

The corresponding vertical density profile of a "Keplerian" disk can be found, if the z-component $k_z = -(\gamma M_c) \cdot z/(z^2 + r^2)^{3/2}$ of the gravitational force of the central mass M_c is taken into account instead that of the gravitational disk potential. This gives

$$\frac{\mathrm{d}}{\mathrm{d}z} \ln \varrho_0 = -\frac{\gamma M_c}{c^2} \frac{z}{(r^2 + z^2)^{3/2}} \tag{5.110a}$$

and consequently with $R^2 = r^2 + z^2$

$$\varrho_0{}^{(K)} = \varrho_0{}^{(K)}(z = 0) \exp \left\{ \frac{\gamma M_c}{c^2} \Big/ R \right\}. \tag{5.110b}$$

This reduces for $z \ll r$ to

$$\varrho_0{}^{(K)} = \varrho_0{}^{(K)}(z > 0) \exp \{z^2/2H_{\text{K}}{}^2\} \tag{5.111}$$

where the scale-height is given with

$$H_{\text{K}} = (c^2/\Omega_{\text{K}}{}^2)^{1/2}. \tag{5.112}$$

It is interesting to note here, that, as a result of the z-dependence of the radial component $k_r = (-\gamma M_c)\, r/(r^2 + z^2)^{3/2}$ of the gravitational force of the central body, the angular frequency is z-dependent. For $|z| \ll r$ follows

$$\Omega^2 = \Omega_{\text{K}}{}^2(r, z = 0) \left(1 - 3\, \frac{|z|}{r}\right). \tag{5.113}$$

This is connected with a vertical shear

$$\frac{\partial v_\varphi}{\partial z} \approx -\frac{3}{2}\, \Omega_{\text{K}}. \tag{5.114}$$

This shear can be a cause for an acceleration of "higher" parts $z \neq 0$ of the disk, if viscosity $\nu \neq 0$ is taken into account. Consequently, "upper" (and more gaseous) parts of the disk should move outward (driven by the angular momentum transport from the Keplerian motion at $z = 0$). This process is most effective in inner regions ($\sim \Omega_{\text{K}}$) and leads to an additional outward transport of gaseous constituents.

The surface mass densities in the above discussed different models for vertical disk profiles are given with

$$\sigma^{(K)} = \sqrt{2\pi}\, \varrho_0(z = 0)\, H_{\text{K}} \tag{5.115}$$

for the "Keplerian disk" and with

$$\sigma^{(g)} = 2\sqrt{2}\, \varrho_0(z = 0)\, H \tag{5.116}$$

for the gravitationally bound disk.

It has to be noted here, that the analytical description of the "Keplerian" disk has the disadvantage, that this model for a ground state of a disk, leads (due to the term $\sim \exp(-z^2/2K_{\text{K}}{}^2)$) to differential equations, which can not be solved analytically. The mathematical structure of the ground state of the gravitationally bound disk is therefore more appropriated to be used as a "standard-model" for the vertical·structure of disks also for those vertical scale heights, which have been determined independently. The numerical differences between the two model-approximations are small, but the mathematical treatment is much easier for models, described with (5.109a) (see the discussion in section 5.3.4.).

5.2.9. Model Parameters of Preplanetary and Presatellite Disks

The following frequencies are characteristic for the physical situation in preplanetary (presatellite) disks, as they have been described in the foregoing sections:

a) $\omega_g^2 = 4\pi\gamma\varrho$ can be used to characterize the time-scale $\tau_g = 2\pi/\omega_g$ of processes, caused by the gravitation of a mass-distribution of density ϱ. This is discussed more in detail in the next chapter. The constant of gravitation is given in this relation with $\gamma = 6.674 \cdot 10^{-11}$ m^3/kg s^2.

b) $\Omega_K^2 = \gamma M_c/a^3$ is the square of the "Kepler frequency" Ω_K of the revolution of a relatively small mass around a massive central body of mass M_c at an ellipse with a long semi-axis "a" (gravitational two-body problem). Circular motions are characterized by $a = R$, where R is the orbital radius. The influence of tidal actions can be described too with this parameter Ω_K (for more details see section 5.3.1.).

c) $\omega_a^2 = c^2k^2$ characterizes acoustic waves with velocity "c" and wave-number "k".

d) $\omega_f^2 = \nu^2/L^4$ is a measure for the square of the inverse characteristic time τ_f for viscosity-caused processes, as they have been discussed in the foregoing sections. Throughout this book, $\nu = 10^{-3}(\gamma M_c R)^{1/2}$ is used according to Lynden-Bell and Pringle (1974). This value indicates that viscosity becomes effective in the large scales in time-scales of the order of 10^3 orbital periods. Consequently, the time-scales for the postulated gravitational structuring processes have to be smaller than 10^3 orbital periods. As will be shown in sections 5.3.2. and 5.3.3. by linear stability analysis, the above mentioned gravitational instability can evolve in preplanetary (presatellite) disks if $\omega_g^2 \gtrsim \Omega_K^2 + \omega_a^2$. The comparison of the rhs-term shows that Ω_K^2 dominates in the large scales (and with $k^2 \approx R^{-2}$) if $M_c > c^2R/\gamma \approx 10^{15}$ R [m]. This is fulfilled in the regions of the planetary system and of the satellite systems (cf. the discussion in section 5.2.4.). This teaches, that indeed local gravitational collapse can overcome the action of tidal forces in (evolving) preplanetary (presatellite) disks, if $\omega_g^2 \gtrsim \Omega_K^2$, or $\varrho(t) \geqq \varrho_c = M_c/4\pi R^3 \approx M_c(t)/R^3$. In principle, this result has been shown even by Gurevitch and Lebedinsky (1950) and by Polyatchenko and Friedmann (1972). This situation was obviously not realized in late protostellar or T-Tauri phases, but it can have been given at earlier phases, when the central body was yet forming, i.e. when the mass of the disk m_d was describable with $m_d \approx 10^{-1}M_c$ (while $H \approx 10^{-1}$ R).
Here it shall be stated more generally, that the large-scale increase of mass density is stopped in regions where $\omega_g^2 \gtrsim \Omega_K^2$. This is due to the local gravitational collapse, feeded by the exceeding mass. Consequently, ϱ_c is a maximum density for disks of massive central bodies.

Based on the results of the foregoing sections, the following conclusions can be drawn for the structure of the preplanetary disk:

— There appears an inner boundary of the disk at $r = r_A$ after the formation of a protostellar magnetic field (see section 5.2.2.), which may have been modified later, when the influence of the magnetic field was not so essential.

— Remanent magnetization of meteoritic inclusions indicates the existence of magnetic fields of (0.1—1) Gauss during the formation of these solids (BRECHER, 1971; BRECHER and ARRHENIUS, 1974). This magnetic field must have had an influence on the structure of the innermost parts of the disk.
— Early thermal heating and radial outflow at greater heights of the disks and later radiation pressure of the electromagnetic radiation cause partially the original gas and volatiles to escape from the inner disk, which consists in its later phases then mainly of chemically fractionated dust and gas. Consequently, inner parts of the outer disk are enriched with gas from the inner disk. This should result in a higher content of volatiles in the outer disk, compared with the original composition.
— A further deviation from original ("solar") composition in the outer disk can have resulted from the inward motion of dust and particles due to dust-gas interactions. Consequently, the outer disk should have had a relatively greater content of volatiles then the original solar nebula.
— The inner disk should have been enriched relatively with chemically fractionated dust and gas.
— The outer radius of the disk is given by equ. (5.59) for $L_p \gtrsim 10^{13}$ m, when the outward directed pressure of the ground state overcomes the gravitation of the central mass.
— The radial dependence of density of the disk is difficult to determine.

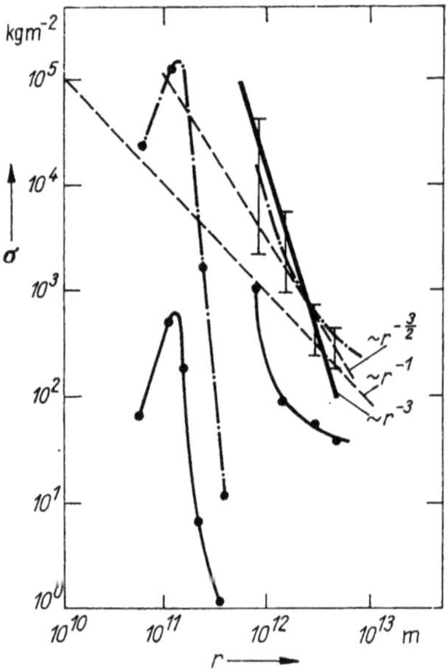

Fig. 5.12. Surface mass densities of the preplanetary disk for the planetary system. The thin solid lines indicate the "minimum mass" model, based directly on the actual planetary masses. The dashed-dotted line describes the values for the "solar composition" model for the inner planets, while the error bars indicate the range of the corresponding "solar composition" models for the outer system (WEIDENSCHILLING, 1977).

Simple estimations were given with equ. (5.32). Actual numerical results (Lin and Papaloizon, 1985) indicate a more or less constant surface density in the inner disk up to about 10^{12} m and a rapidly decreasing surface density for greater distances. But these estimations are very crude. They have to be precised by data as they can be derived from the real mass distributions in the four satellite systems of the Sun, Jupiter, Saturn and Uranus.

The data about these systems have been interpreted within the frame of two limiting models. The "minimum mass model" is based on the assumption that nearly all of the preplanetary disk matter accumulated to planets, while the "solar composition model" is constructed by adding the solar complement of light elements to each planet (Weidenschilling, 1977), to reconstruct protoplanets of an originally solar composition. As has been discussed above, the solar composition model should give to low values for the density in the outer system. This is due to the higher content of volatiles, compared with that of the original ("solar") composition.

Now, the results of these comparisons with real radial mass distributions and derived radial density profiles for different models shall be discussed (see Tables 5.6, 5.7 and 5.8).

The surface mass densities $\sigma(r)$, as they have been described by the Tables 5.6 to 5.8 are shown in Figs. 5.12 and 5.13 It is interesting to note, that the radial surface density profiles differ essentially in all the four systems. Having in mind the radial regularities or quasi-periodicities of orbits in these systems, this discrepancy may indicate, either that details in the radial structure of the preplanetary disk do not influence essentially its further evolution, or that the early surface density of the preplanetary disk was quite different

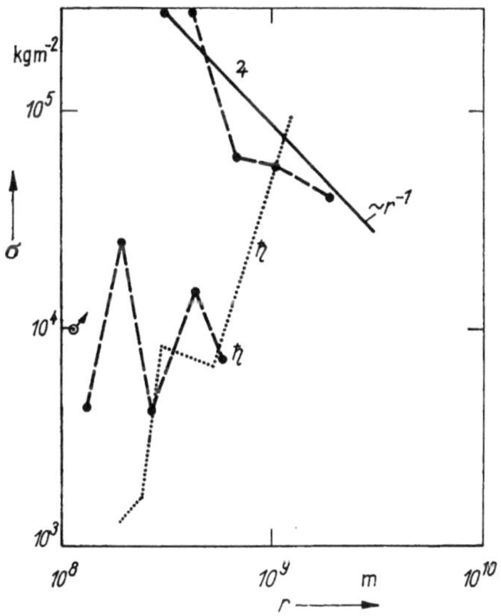

Fig. 5.13. Surface mass densities for "minimum mass" models for the satellite system of Jupiter, Saturn and Uranus.

Table 5.6. Minimum mass model of the planetary system

Planet	Semimajor axis (10^{11} m)	n^2 (s^{-2})	r_i (10^{11} m)	r_a (10^{11} m)	T (10^2 k)	σ_{min} (kg/m²)	c^2 (10^6 m²/s²)	$H_r^{(k)}$ (10^{10} m)	$\varrho_{00}^{(k)}$ (10^{-9} kg/m³)
Mercury	0.579	$6.53 \cdot 10^{-13}$	0.3	0.73	6.33	24	3.74	0.28	4.84
Venus	1.08	$1.01 \cdot 10^{-13}$	0.73	1.29	4.63	137	2.74	0.60	12.9
Earth	1.50	$3.76 \cdot 10^{-14}$	1.29	1.89	3.93	100	2.32	0.91	6.20
Mars	2.28	$1.07 \cdot 10^{-14}$	1.89	3.0	3.19	4	1.89	1.53	0.15
Asteroids	3.7	$2.50 \cdot 10^{-15}$	3.0	5.7	2.50	—	1.48	2.81	—
Jupiter	7.78	$2.69 \cdot 10^{-16}$	5.7	11.0	1.73	683	1.02	7.11	5.42
Saturn	14.3	$4.34 \cdot 10^{-17}$	11.0	21.5	1.27	53	0.753	15.2	0.20
Uranus	28.7	$5.36 \cdot 10^{-18}$	21.5	36.9	0.90	3	0.530	36.3	0.005
Neptune	45.0	$1.39 \cdot 10^{-18}$	36.9	52.0	0.72	2	0.425	63.8	0.002
Pluto/ Charon	59.5	$6.17 \cdot 10^{-19}$	52.0	66.0	0.63		0.371	89.5	

It has to be noted here, that within the frame of this model the vertical scale $H_v^{(g)}$ is greater by three orders of magnitude or more than $H_r^{(k)}$. This indicates that disk-gravitation for this model is not essential for the vertical mass distribution. Here: $c^2 = 3kT/2\bar{m}_{H_2}$ with $Tr^{1/2}$ = const. and a solar surface temperature of 5770 K. The validity of $Tr^{1/2}$ = const. is restricted to optically thin (and consequently late) preplanetary conditions. Otherwise higher temperatures have to be used in the inner parts while the outer parts were cooler.

Table 5.7. Solar composition model of the planetary system

Planet	Surface density* (10^3 kg/m^2)	$H_v^{(g)}$ (10^{10} m)	$H_v^{(k)}$** (10^{10} m)	$\varrho_0^{(k)}$ (10^{-8} kg m^{-3})
Mercury	8.8	70	0.28	177
Venus	47.5	10	0.60	447
Earth	32.0	12	0.91	198
Mars	0.95	324	1.53	3.5
Asteroids	0.013–0.4	18630–608	2.81	0.03–0.8
Jupiter	1.2–24	140–7	7.11	1.0–20
Saturn	0.55–3.3	220–38	15.2	0.2–1.2
Uranus	0.15–0.4	560–225	36.3	0.02–0.06
Neptune	0.1–0.25	675–270	63.8	0.009–0.02

* according to WEIDENSCHILLING (1977)
** from Table 5.6

Table 5.8. Minimum mass model of the satellite systems of Jupiter, Saturn and Uranus

Satellites	Semimajor axis (10^8 m)	r_i (10^8 m)	r_a (10^8 m)	m (10^{21} kg)	σ_{min} (10^3 kg/m^2)	$\varrho_0^{(k)}$ (10^{-5} kg/m^3)
Io	4.216	3.6	5.5	89.2	164	328
Europa	6.709	5.5	8.7	48.6	34	34
Ganymede	10.70	8.7	14.8	148.9	33	16
Callisto	18.80	14.8	23	106.3	11	2.3
Mimos	1.86	1.70	2.12	0.037	0.73	2.75
Enceladus	2.38	2.12	2.70	0.084	0.96	2.49
Tethys	2.95	2.70	3.40	0.630	4.7	8.84
Dione	3.77	3.40	4.50	1.2	4.4	5.72
Rhea	5.27	4.50	8.75	2.2	1.2	0.96
Titan	12.22	8.75	13.5	140	42	9.37
Hyperion	14.81	13.5	18			
Miranda	1.30	1.14	1.61	0.014*	0.45	1.13
Ariel	1.92	1.61	2.30	0.27*	2.9	4.06
Umbriel	2.67	2.30	3.53	0.10*	0.44	0.38
Titania	4.38	3.53	5.12	0.52*	1.2	0.52
Oberon	5.86	5.12	8	0.38*	0.32	0.08

* The masses have been determined from radii of 150 km, 400 km, 275 km, 500 km and 450 km respectively and with an assumed mass density of the satellites of 10^3 kg/m^3.

from that, which has been reconstructed from actual planetary (and satellite) masses or models of their protoplanets. The last mentioned possibility seems to be more realistic. This implies that most of the preplanetary and presatellite matter has been dispersed during planetary growth. It should have been fallen mainly onto the Sun or moved out of the planetary system.

5.3. Evolution of Preplanetary Disks and Rings

The foregoing chapter has shown, that around a yet forming protostar (protoplanet) a protostellar disk evolves which should be the site of the preplanetary planetogonic processes. During these early phases, collective gravitational processes could be able to structurate the disk in that way as it has been inferred from the quasiperiodic structure in the above discussed four systems. An advantage of this preplanetary structuring mechanism is its analogous applicability to the disks of all these four systems. This possible planetogonic mechanism will be studied more in detail in this chapter.

5.3.1. Local Gravitational Collapse and Tidal Actions

Local fluctuations in the mass density may accumulate more mass due to their selfgravitation. In principle, this could be a mechanism for the origin of planets from direct gravitational instabilities in the circumstellar envelope or the preplanetary disk. But to be effective, selfgravitation has to overcome the action of the tidal force of the central mass M_c. This can be estimated by comparing the gravitational force density k_g (of the cloud of preenhanced density ϱ_{crit} within a radius R)

$$k_g = \gamma \frac{4}{3} \pi R^3 \varrho_{\text{crit}} / R^2 = \frac{4}{3} \pi \gamma R \varrho_{\text{crit}} \tag{5.117}$$

with the differential gravitation $\mathrm{d}k$ (tidal action) of the central mass M_c acting upon the extended cloud at the surface R with its mass center at $r = r_c$ moving on a Keplerian orbit $v_\varphi^2/r_c = -\dfrac{\gamma M_c}{r_c^2}$.

$$|\mathrm{d}k| = \frac{2\gamma M_c}{r_c^3} \cdot R. \tag{5.118}$$

This gives as a condition for the collapse to proceed

$$\varrho_{\text{crit}} > \frac{3M_c}{2\pi r_c^3}. \tag{5.119}$$

A more detailed derivation of this relation giving, $\varrho_{cr} = M/4\pi r^3$ will be given in the next chapters. Numerically follows for $r = 1$ AU $= 1.495 \cdot 10^{11}$ m and $M_c = 1.99 \cdot 10^{30}$ kg (actual solar mass)

$$\varrho_c \approx 3 \cdot 10^{-4} \text{ kg m}^{-3}. \tag{5.120}$$

In all other cases tidal action disrupts the accumulating cloud, preventing on this way any structuration in the density. It is interesting to note here, that density distributions similar to equ. (5.119) between an inner radius r_i and outer radius r_a of a disk with $r_a \gg r_i$ are connected with a total mass m_d of this disk of height H

$$m_d = 2\pi H \int dr r \varrho_c, \tag{5.121}$$

$$m_{\text{disk}} \approx \frac{3H}{r_i} M_c \approx 10^{-1} M_c, \tag{5.122}$$

when the vertical height of a selfgravitating disk with $H < 10^{-1} r$ is used (cf. equation 5.109b)).

Consequently, structuring gravitational instabilities in a disk are possible only, when the mass of the disk is of the order of or greater than some tenth of the central mass. This is an indication, that this kind of structuring processes can have been working only in early phases when the central body was yet forming (M_c is yet relatively small and the inward flow of mass of the disk is decelerated by the action of viscosity).

In the same way the density can be estimated which is necessary for self-gravitation to overcome the action of pressure. The force density of pressure, acting on a scale R is given by

$$k_p \approx \frac{1}{\varrho} \frac{c^2}{R} \varrho \approx \frac{c^2}{R}. \tag{5.123}$$

This gives with equ. (5.112) the condition for dominating selfgravity at these scales

$$\frac{4}{3} \pi \gamma R \varrho_{\text{crit}} \gtrsim \frac{c^2}{R} \quad \text{or} \quad \varrho_{\text{crit}} \gtrsim \frac{3c^2}{4\pi\gamma R^2} \tag{5.124}$$

with $\omega_g^2 = 4\pi\gamma\varrho_{\text{crit}}$ this takes the form

$$\omega_G^2 \gtrsim \frac{c^2}{R^2}. \tag{5.125}$$

Equations (5.119) and (5.120) have been derived here by simple estimations to make the physics behind these equations more plausible. These relations will be derived now more in detail on an exact mathematical basis. The results of these more sophisticated calculations can be understand then simply on the basis of the above given results.

5.3.2. Local Dispersion Relations

To investigate more exactly local stability of the ground state of the gaseous disk against small perturbations, the relevant equations shall be formulated for a coordinate system corotating (not comoving!) with the a constant angular velocity $\mathbf{\Omega} = \text{const}$. Then the equation of motion takes the form

$$\frac{\partial \boldsymbol{v}}{\partial t} + (\boldsymbol{v} \operatorname{grad}) \boldsymbol{v} + 2\mathbf{\Omega} \times \boldsymbol{v} + \mathbf{\Omega} \times (\mathbf{\Omega} \times \boldsymbol{r}) = -\operatorname{grad} (V + \Phi) - \frac{1}{\varrho} \operatorname{grad} p . \tag{5.126}$$

Neglecting the small influence of Φ_0 and p_0 in large scales (compared with $V = -\gamma Mc/r$), it follows for a stationary axisymmetric "ground state" in cylindrical coordinates with $\boldsymbol{v}_0 = \boldsymbol{e}_\varphi \boldsymbol{v}_0(r)$

$$v_0(r) = \pm \left(\gamma \frac{Mc}{r}\right)^{1/2} - \Omega r. \tag{5.127}$$

The signs $\pm$ indicate both the possible orientations of Keplerian rotation (prograde, retrograde). With the choice $\mathbf{\Omega} \equiv \mathbf{\Omega}_\text{K}(r = R) = (\gamma Mc/R^3)^{1/2}$ follows with $x = r - R$ for the prograde orientation (positive sign in equ. (5.127)) with Taylor expansion

$$v_0(x) = -\frac{3}{2} \left(\gamma \frac{Mc}{r^3}\right)^{1/2} x = -\frac{3}{2} \Omega_\text{K}(r = R)\, x \tag{5.128}$$

(for retrograde motion follows $v_0(x) = -2\Omega_\text{K}(R)\, R - \frac{1}{2} \Omega_\text{K}(R)\, x$). Linearization of the equations of motion and continuity gives with $\boldsymbol{v} = \boldsymbol{v}_0 + \boldsymbol{v}$, $\varrho = \varrho_0 + \delta\varrho$, $p = p_0 + \delta p$, $\Phi = \Phi_0 + \Psi$, $\varrho_0 = \text{const.}$, $p_0 = \text{const.}$ $\delta_p = c^2 \delta\varrho$, $(c^2 = \bar{\gamma} kT/\mu m_H)$ for adiabatic processes in local corotating Cartesian coordinates $x = r - R$, $y = R_\varphi$ at $r = R$ and for prograde motion

$$\frac{\partial}{\partial t} \delta\varrho + \varrho_0 \frac{\partial}{\partial x} U_x + \varrho_0 \frac{\partial}{\partial y} U_y = 0, \tag{5.129a}$$

$$\frac{c^2}{\varrho_0} \frac{\partial}{\partial x} \delta\varrho + \frac{\partial}{\partial t} U_x - 2\Omega_K U_y + \frac{\partial \Psi}{\partial x} = 0, \tag{5.129b}$$

$$\frac{c^2}{\varrho_0} \frac{\partial}{\partial y} \delta\varrho + \frac{1}{2} \Omega_K U_x + \frac{\partial}{\partial t} U_y + \frac{\partial \Psi}{\partial y} = 0, \tag{5.129c}$$

$$4\pi\gamma\delta\varrho - \Delta\Psi = 0. \tag{5.129d}$$

For locally constant temperature and density, and due to the restriction on the neighbourhood of $r = R$ and resulting $\mathbf{\Omega} = \mathbf{\Omega}_\text{K}(R) = \text{const.}$, the conditions for solutions $\sim \exp(\omega t + ik_x x + ik_y y)$ can be found from this set of homogenous equations. For the special model of an exact two-dimensional disk with mass density $\delta\varrho \sim \sigma\delta(z)$, where $\delta(z)$ is the Dirac delta function, equ. (5.129d) can be replaced by $\left.\frac{\partial \Psi}{\partial z}\right|_{z \to +0} = 2\pi\gamma\delta\sigma$, where $\delta\sigma$ is the variation of the surface density and Ψ the corresponding gravitational potential. As will be discussed more in detail in 5.3.4., this result can be derived by integrating (5.129d) over z. With $\delta\varrho$ (and $\delta\sigma$), U_x, U_y, $U_z = 0$ and $\Psi \sim \exp(\omega t + ik_x x + ik_y y)$

follows the dispersion relation (3-dimensional case)

$$\omega^3 + \omega\{\Omega_K^2 + c^2(k_x^2 + k_y^2) - \omega_g^2\} + \frac{3}{2}\Omega_K \frac{k_x k_y}{k_x^2 + k_y^2} [c^2(k_x^2 + k_y^2) - \omega_g^2] = 0 \tag{5.130}$$

where $4\pi\gamma\varrho_0 = \omega_g^2$ has been introduced.

For the above mentioned two-dimensional models follows the corresponding dispersion relation

$$\omega^3 + \omega\left[(k_x^2 + k_y^2)\left(c^2 - \frac{2\pi\gamma\sigma_0}{k_z}\right) + \Omega_K^2\right] + \frac{3}{2}\Omega_K k_x k_y \left[c^2 - \frac{2\pi\gamma\sigma_0}{k_z}\right] = 0, \tag{5.131}$$

where k_z follows via $\left.\frac{\partial\Psi}{\partial z}\right|_{z\to+0} = -k_z\Psi(z = 0)$. This k_z^2 is fixed by $\Delta\Psi = 0$ for $z \neq 0$, leading to $-k_z^2 = k^2 = k_x^2 + k_y^2$. These relations reduce for axi-symmetric conditions ($k_y = 0$) to the well known relations

$$\omega^2 = \omega_G^2 - \Omega_K^2 - c^2k_x^2, \tag{5.132}$$

and

$$\omega^2 = \frac{k_x^2}{k_z} 2\pi\gamma\sigma_0 - \Omega_K^2 - c^2k_x^2 \tag{5.133}$$

respectively. They shall be discussed now. Equations (5.132) and (5.133) indicate, that there can indeed evolve gravitational instabilities if $\omega^2 > 0$ or $\omega_g^2 > \Omega_K^2 + c^2k^2$ or $2\pi\gamma\sigma_0 k_x > \Omega_K^2 + c^2k_x^2$ respectively. The character of the solutions changes at $\omega = 0$. This so called "marginal instability" shall be investigated now more in detail and in comparison with parameters for preplanetary disks, as they have been derived in 5.2.8. for the mass densities.

From equ. (5.132) it can be seen that pressure stabilizes small scale disturbances. On the other side, large-scale disturbances can be stabilized by Keplerian motion for $\Omega_K^2 \gg \omega_g^2$ with $\omega \simeq \pm i\Omega_K$. But is every case, there is a range of instability, especially in the longer scales, if the mass density of the disk is sufficiently high.

The corresponding relation $\omega_g^2 = 4\pi\gamma\varrho_0 \gtrsim \Omega_K^2 = \gamma Mc/r^3$ is, up to a numerical factor equivalent to equ. (5.119), indicating the local dominance of self-gravitation (of the disturbance) over the radially disruptive tidal forces from the central body. This situation has been discussed also in 5.3.1. It will be shown later, that this condition can have been realized at early protoplanetary conditions.

By numerical comparison it can be seen very simply, that also the condition $\omega_g^2 > c^2k^2$ (cf. 5.124) can have been fulfilled at early preplanetary conditions. Using mass-densities as they were given with Table 5.7. it can be seen that 10^{-6} kg m$^{-3} \leq \varrho \lesssim 10^{-10}$ kg m^{-3} for the solar composition model. Consequently follows 10^{-15} s$^{-2} \lesssim \omega_g^2 \lesssim 10^{-19}$ s^{-2}. The orders of magnitude for the square $\Omega_K^2 = \gamma \frac{M_c}{r^3}$ of the Keplerian angular velocity can be shown to be greater by one order of magnitude or more in the corresponding regions.

Consequently, the disk mass density has to be greater by this factor to have appropriated conditions for gravitationally structuring processes to occur in an (eventually early) preplanetary disk.

5.3.3. Azimutal Instability of Rings

It has been shown in the foregoing section that for axisymmetric conditions a gravitationally caused radial structuring (by instability) of a disk is possible, and that the conditions for this instability could have been given in the preplanetary disk. Here, the structuring process should have been acting in that way that the disk has been separated into rings (of enhanced density). The next step is to investigate the further evolution of these rings. Are they a stable formation, where a planetary growth resulted then from stochastic accretion of planetesimals, or are they unstable on a shorter time-scale, triggering then a faster planetary growth?

This question can be answered by investigating the dispersion relations (5.130) und (5.131). These equations can by unified with $f = \omega/\Omega_K$ and

$$g^2 = [\omega_g^2 - c^2(k_x^2 + k_y^2)]/\Omega_K^2 \tag{5.134}$$

and

$$g^2 = (k_x^2 + k_y^2)\,(2\pi\gamma\sigma_0/k_z - c^2)/\Omega_K^2 \tag{5.135}$$

for the 3D- and the 2D-case respectively. Then they take the form of the cubic equation

$$f^3 + f(1 - g^2) - \frac{3}{2}\, g^2 \frac{k_x k_y}{k_x^2 + k_y^2} = 0\,. \tag{5.136}$$

Equations of this type can be solved analytically by Cardani's formula, giving for equations of the type $y^3 + py + q = 0$ the solutions $y_1 = u + v$, $y_2 = -(u + v)/2 + \mathrm{i}\sqrt{3}(u - v)/2$, $y_3 = -(u + v)/2 - \mathrm{i}\sqrt{3}(u - v)/2$, where $u = \left(-q/2 + \sqrt{D}\right)^{1/3}$, $v = \left(-q/2 - \sqrt{D}\right)^{1/3}$ and $D = (p/3)^3 + (q/2)^2$.

For $D > 0$ there exist one real and two conjugated complex solutions, while three real solutions can be found for $D < 0$.

Without solving these equations in detail, it can be seen that there open new possibilities for instabilities to occur. The geometric factor $3k_xk_y/2k^2$ leads with $k_y = m/2\pi R$ and $k_x = lH^{-1}$ (when the internal structure of the ring is assumed to be governed by its gravitation, analogous to equ. (5.108)) to

$$\frac{3}{2}\frac{k_xk_y}{k^2} \approx \frac{3mH}{4\pi Rl} \approx 10^{-1}\text{—}10^{-2}\,. \tag{5.137}$$

The condition for marginal instability (i.e. $\omega \to 0$) was given for the axisymmetric case with $g^2 \to 1$ (when the effects of pressure are neglected). Under these conditions follows now with $g^2 \to 1$ that $p \to 0$ in Cardinis formula. With $g = -3k_x k_y / 2k^2$ follow then approximately $D = |g/2|$, $u = g^{1/3}$, $v = 0$, and consequently

$$\omega_1 = -\Omega_K \left(\frac{3}{2} \frac{k_x k_y}{k^2}\right)^{1/3} \approx -\frac{\Omega_K}{5} \tag{5.138}$$

$$\omega_2 = 10^{-1}\,\Omega_K - i\,\frac{\sqrt{3}}{2} \cdot 10^{-1}\,\Omega_K \tag{5.139}$$

$$\omega_3 = 10^{-1}\,\Omega_K + i\,\frac{\sqrt{3}}{2} \cdot 10^{-1}\,\Omega_K\,. \tag{5.140}$$

The essential result is that there exist solutions with a positive argument, i.e. increasing exponentially with time. The characteristic time for these azimutal instabilities to develope is about 10 times greater than that for radial structuring ($\omega_g \approx \Omega_K$). This is the necessary condition, to describe the radially structuring gravitational processes in an axisymmetric approximation.

The following additional conclusions and remarks can be made in connection with the new type of solutions of the cubic equation (5.136), compared with that of the quadratic equations for the axisymmetric case:

- For dominating pressure ($g^2 < 0$) sound waves can evolve in the differentially rotating disk, having exponentially increasing amplitudes. They could dissipate in the non-linear regime and be stabilized on a level of non-linear turbulence. This supports the possibility of turbulence and related processes of transport and particle growth in the very early proeplanetary disk, as they have been discussed by Völk (1984).
- The condition $g^2 < 1$ for positive g^2, i.e. for dominance of selfgravitation over pressure, opens a principially new possibility for planetary growth from the transient state of preplanetary rings. This is an alternative approach to planetogony, compared with that of a stochastic growth by collisional accretion. This result describes an essential step in the proposed planetogonic scenario.

5.3.4. Gravitational Instabilities of Disks

It has been shown in the foregoing sections that the dispersive action of pressure in the preplanetary disk can be neglected for processes with large radial scales of the order $\gtrsim 10^{-1} r$, where r is the distance from the central body. In this case, the dynamics of a (preplanetary) disk above a central body and interesting only via (collective) selfgravity can be described by

$$\frac{\partial}{\partial t}\,\boldsymbol{v} + (\boldsymbol{v}\,\mathrm{grad})\,\boldsymbol{v} = -\mathrm{grad}\,(V + \Phi)\,, \tag{5.141}$$

$$\Delta\Phi = 4\pi\gamma\varrho \tag{5.142}$$

which together with the equation of continuity completes a closed set of equations. To investigate the stability of the "ground state", as it has been discussed in 5.3.2. with v_0, ϱ_0, $V = \gamma M_c/r$ and Φ_0, a linearization according to $\boldsymbol{v} = \boldsymbol{v}_0 + \boldsymbol{u}$, $\varrho = \varrho_0 + \delta\varrho$, $\Phi = \Phi_0 + \Psi$ with $|\boldsymbol{u}| \ll |\boldsymbol{v}_0|$ and $\delta\varrho \ll \varrho_0$ gives at $z = 0$

$$\frac{\partial}{\partial t}\boldsymbol{u} + (\boldsymbol{v}_0 \operatorname{grad})\,\boldsymbol{u} + (\boldsymbol{u} \operatorname{grad})\,\boldsymbol{v}_0 = -\operatorname{grad}\Psi \tag{5.143}$$

$$\Delta\Psi = 4\pi\gamma\delta\varrho. \tag{5.144}$$

Using furthermore the approximation that the "horizontal" contribution of Φ_0 can be neglected compared with that of the central mass, v_0 reduces to that of the undisturbed Keplerian motion $v_0 = (\gamma M_c/r)^{1/2}$, defining with $v_0 = \Omega_K \cdot r$ the Keplerian angular velocity $\Omega_K = (\gamma M_c/r^3)^{1/2}$.

Then equ. (5.141) can be reduced in the axisymmetric case to

$$\frac{\partial U_r}{\partial t} - 2\Omega_K U_\varphi = -\frac{\partial}{\partial r}\Psi \tag{5.145a}$$

$$\frac{1}{2}\Omega_K U_r + \frac{\partial U_\varphi}{\partial t} = 0. \tag{5.145b}$$

With $\frac{\partial}{\partial t} = \omega$ it follows then for the dependence of the velocities on the gravitational potential:

$$U_r = -D^{-1}\omega\,\frac{\partial\Psi}{\partial r} \tag{5.146a}$$

$$U_\varphi = D^{-1}\,\frac{\Omega_K}{2}\,\frac{\partial\Psi}{\partial r} \tag{5.146b}$$

where $D = \omega^2 + \Omega_K{}^2$ has been used.

It can be seen very simply, that these equations are able to describe a growth of rings for appropriated conditions by an instability ($\omega > 0$, real). In this case follows $U_r = -(\omega/D)\,\frac{\partial\Psi}{\partial r}$, giving in the neighbourhood of a massive ring (local minimum of Ψ) at $r = R$ that $\frac{\partial\Psi}{\partial r} < R$ for $r < R$, and consequently $v_r > 0$ at $r < R$. In the same way it follows $U_R < 0$ for $r > R$. Consequently, there is a mass flow towards $r = R$ from both sides. This is equivalent to a growth of the ring as long as the linear approximation can be used.

Equations (5.145) can be used now to introduce v_r and v_φ into the linearized equation of continuity

$$\varrho_0\left(\frac{1-n}{r} + \frac{\partial}{\partial r}\right)U_r + \frac{\partial}{\partial t}\,\delta\varrho = 0, \tag{5.147}$$

where $U_z = 0$ and $\varrho = c_0 r^{-n}$ have been introduced.

This gives

$$\frac{\partial^2}{\partial r^2}\Psi + \frac{4-n}{r}\,\frac{\partial\Psi}{\partial r} - \frac{\Omega_K{}^2}{4\pi\gamma\varrho}\,\Delta\Psi = 0. \tag{5.148}$$

Taking into account now that $H/r < 1$ (see equ. (5.109b)) for a finite thin disk with a vertical structure determined by its gravitation, it follows for large-scale radial structures instead of (5.144)

$$4\pi\gamma\delta\varrho = \Delta\Psi \approx \frac{\partial^2}{\partial z^2}\Psi, \tag{5.149}$$

when z is the vertical coordinate. Now the potential Ψ can be determined from

$$\frac{\partial^2}{\partial r^2}\Psi + \frac{4-n}{r}\frac{\partial}{\partial r}\Psi - \frac{M_c}{4\pi c_0} r^{n-3}\frac{\partial^2\Psi}{\partial z^2} = 0. \tag{5.150}$$

Taking into account now the z-dependence of the (undisturbed) density of a "gravitationally bound" disk as it is described with equ. (5.109a), it follows with complete representation for this "standard model" (see the discussion at the end of section 5.2.8.)

$$\varrho(r, z) = \frac{Kr^{-n}}{\cosh^2\left(z/\sqrt{2}\,H\right)} \tag{5.151}$$

(or $c_0 = K\cos\mathrm{h}^2\left(z/\sqrt{2}\,H\right)$ equivalently, that

$$-\frac{M_c}{4\pi c_0} r^{n-3}\frac{\partial^2\Psi}{\partial z^2} = -\frac{M_c}{4\pi K}\cosh^2\left(\frac{z}{\sqrt{2}\,H}\right) r^{n-3}\frac{\partial^2\Psi}{\partial z^2}. \tag{5.152}$$

Having in mind now, that only radially quasiperiodic solutions of equ. (5.150) are of planetogonic interest, and that such type of solution exists, if

$$\cosh^2\left(\frac{z}{\sqrt{2}\,H}\right)\frac{\partial^2}{\partial z^2}\Psi = -k^2\Psi, \quad k^2 = \text{const.}, \tag{5.153}$$

it follows, that the vertical structure of these solutions has to be of the form

$$\Psi(z) \sim \tanh\frac{|z|}{\sqrt{2}\,H}, \tag{5.154}$$

if "$\pm z$-symmetry" is taken into account. The constant parameter k^2 is fixed with (5.153) and (5.154) to $k^2 = (2H^2)^{-1}$.

From the point of view of physics, the remarkable result of equs. (5.153) and (5.154) is, that there form mass accumulations at $z = 0$ in the midplane of the disk (where the undisturbed density has its maximum value). For the surface density $\delta\sigma(r)$ follows from $\delta\varrho = \delta\sigma\delta(r)$ after z-integration of $\Delta\Psi \approx \frac{\partial^2}{\partial r^2}\Psi = 4\pi\gamma\delta\varrho$ that

$$\delta\sigma(r) = (4\pi\gamma)^{-1}\left(-2\left.\frac{\partial\Psi}{\partial z}\right|_{z\to+0}\right), \tag{5.155}$$

where $\int\limits_{-\infty}^{\infty} \mathrm{d}z\,\frac{\mathrm{d}^2\Psi}{\mathrm{d}z^2} = -\left.\frac{\mathrm{d}\Psi}{\mathrm{d}z}\right|_{z\to+0} + \left.\frac{\mathrm{d}\Psi}{\mathrm{d}z}\right|_{z\to-0} = -2\left.\frac{\mathrm{d}\Psi}{\mathrm{d}z}\right|_{z\to+0}$.

With

$$\Psi(r, z) = f(r) \tanh \left(|z|/\sqrt{2}\, H\right) \tag{5.156}$$

follows

$$\delta\sigma(r) = -\frac{1}{4\pi\gamma H \sqrt{2}} f(r) . \tag{5.157}$$

The radial distribution $f(r)$ has to be determined from the remaining "radial potential equation", as it follows by introducing equ. (5.156) into (5.150).

This gives

$$\frac{d^2}{dr^2} f + \frac{4 - n}{r} \frac{d}{dr} f + \frac{M_c}{8\pi K H^2} r^{n-3} f(r) = 0 . \tag{5.158}$$

Solutions of this differential equation are given with $H^2 = H_0{}^2 r^m$ in the form

$$f(r) = C_1 r^{-\alpha} J_p(\varkappa r^\beta) + C_2 r^{-\alpha} J_p(\varkappa r^\beta) , \tag{5.159}$$

where J_p are Bessel's functions of first kind and p-th order, and where the parameter-coefficients are determined by

$$\alpha = \frac{3 - n}{2} \tag{5.160a}$$

$$\beta = \frac{n - m - 1}{2} \tag{5.160b}$$

$$p = \frac{3 - n}{|n - m - 1|} \tag{5.160c}$$

$$\varkappa = \left(\frac{M_c}{2\pi K H_0{}^2}\right)^{1/2} \Big/ |n - m - 1| \tag{5.160d}$$

C_1 and C_2 are yet arbitrary constants (Smirnov, 1961).

Taking into account now the quasiperiodic character of Bessel's functions (i.e. that the local maxima — and minima and zeros — of $Jp(x)$ are equally spaced by about 2π respectively, i.e. for $x \to x + 2\pi$, where $x = \varkappa r^\beta$). Consequently, there evolve in the linear regime ringlike disturbances $\delta\sigma \sim -f(r)$. If the density is sufficiently high, these rings can grow further by non-linear processes. In other words, the disk has been separated into a system of single (separated) rings or bands. According to the proposed "planetogonic scenario", these rings of enhanced density are assumed to be the building sites of planets (satellites).

Here it has to be mentioned again, that the above derived results were based on a "mathematical structure" of the vertical profiles of the disk, as it has been derived in section 5.2.8. for a "gravitationally bound" disk. This "standard model" was used mainly, to have an analytical solution of equ. (5.153) (which is not integrable if the profile $\varrho_0 \sim \exp(-z^2/2H_K{}^2)$ is used instead of $\cosh^{-2}(z/2H)$). But it shall be shown now, that equivalent results follow for the Keplerian disk. Instead of equ. (5.153), now

$$\frac{\partial^2}{\partial z^2} \Psi = -k^2 \Psi \exp(-z^2/2H_K{}^2) \tag{5.161}$$

has to be solved. An analytical solution can be found for $z \ll H_K$, where $\exp(-z^2/2H_K^2) \approx 1 - z^2/2H_K^2$. Then the solution takes the form

$$\Psi \sim \exp(-z^2/4H_K^2) \tag{5.162a}$$

and

$$k^2 = 1/2H_K^2. \tag{5.162b}$$

Consequently, for $z \ll H_K$ follows from equ. (5.150) with $\Psi \sim f(r) \exp(-z^2/4H_K^2)$ that

$$\frac{d^2}{dr^2} f + \frac{4-n}{r} \frac{d}{dr} f + \frac{M_c}{8\pi K H_K^2} r^{n-3} f(r) = 0. \tag{5.163}$$

This is equivalent to (5.158), where only H_K has to be used instead of H. This result is a further justification for using the "gravitationally bound" disk as a standard model for the vertical structure of preplanetary disks also for independently determined scale heights H. The qualitative character of the derived results remains unchanged. Only small numerical deviations can occur.

It is a remarkable matter of fact of the above given theory that the vertical structure of the relatively thin preplanetary (presatellite) disk has an influence on the character of the radially structuring processes.

It is interesting to note, that in the special case of $\varrho \sim r^{-3}$ and with $H^2 =$ const. this disk can be described in a two-dimensional approximation with $\Delta\Psi = 0$ at $z \neq 0$. This results from the fact, that the character of the radial structures ($\sim J_0(\varkappa r)$ and $Y_0(\varkappa r)$) remains unchanged after z-integration of equ. (5.150), and with $\delta\varrho \to \delta\sigma$ and $\varrho \to \sigma$. Only these resulting solutions $\Psi \sim e^{k'z}\big(C_1 Y_0(k'r) + C_2 J_0(k'r)\big)$ are compatible with $\Delta\Psi = 0$ as it has to be used now for $z \neq 0$.

It has been mentioned by Rüdiger and Tschäpe (1987), that these $2D$-solutions, which in the general case have to satisfy the condition for marginal instability in axisymmetric infinitely thin disks

$$2\pi\gamma\sigma k' - \Omega_K^2 - c^2(k')^2 = 0 \tag{5.164}$$

as it follows, from equ. (5.133) and $k_x = k_z = k'$, can have radially periodic properties (i.e. real k') only, if (as an additional condition for temperature)

$$\frac{\pi\gamma\sigma}{\Omega_K c} \geqq 1 \quad \text{or} \quad T \leqq \frac{\pi^2\gamma^2\sigma^2 m_{Gas}}{\Omega_K^2 \bar{\gamma} k} \tag{5.165}$$

is fulfilled ($c^2 = \bar{\gamma}kT/m_{Gas}$). But, taking into account here that $\sigma \sim c \sim T^{1/2}$, as it follows from equs. (5.109b) and (5.116b), it can be seen that this condition can be reformulated in a temperature-independent and well-known form $(2\pi\gamma K)^{1/2}/\Omega_K = \omega_g/\Omega_K \sqrt{2} \geqq 1$. Consequently, equ. (5.165) is not an additional constraint to the consistent approach, as it has been described above.

It shall be noted here briefly that the above given discussion of the radial stability of a disk took into account explicitely the height-profile of a self-gravitating disk and the r-dependent (differential) Keplerian rotation of the disk. Earlier discussions of this stability of thin disks have been made by Gurevitch and Lebedinski (1950), Polyatchenko and Fridman (1972),

HUNTER (1965) and TOOMRE (1964). HUNTER made use of spheroidal coordinates and Legendre functions. He obtained the frequencies of the free oscillations as eigen-values of an infinite matrix. TOOMRE applied Fourier-Bessel analysis to obtain equilibrium figures for galactic disks, replacing the disk by a finite number of concentric rings for a radially local stability analysis for single rings. Unfourtunately, no attempt was made, to apply these results consequently to the planetogony of the planetary system. The following discussion will demonstrate that the results are indeed compatible with the observed properties of the planetary system and its evolved satellite systems.

5.3.5. Comparison with Real Systems (Numerical Results)

The foregoing two sections gave the theoretical and modelbackground to understand the origin of the Solar system and of the satellite systems of Jupiter, Saturn and Uranus within the frame of a planetogonic process-sequence: preplanetary disk, structuring of this disk into rings via radial gravitational instability, decay of the transient ring-phenomenon into protoplanets by the "azimutal instability".

As a yet open problem remained the exact time t_p of the action of these planetogonic processes. Following MORFILL (1985) and to have a numerical basis for the following discussion, it shall be assumed, that t_p is fixed by the condition, that the central mass $M_c(t_p) \approx \frac{1}{2} M_c$ (final), i.e., that the planetogonic era started while about half of the final mass was accumulated in the yet growing central body.

Here it should be mentioned furthermore, that at the time t_p when $M_c(t_p) \approx 0.5 M_\odot$, there evolves a convection zone in the growing protostar (see chapter 5.1.), which should be connected with a dynamo-generated circumprotostellar magnetic field. The resulting protostellar magnetosphere should have been effective as a "barrier" for the inflow of matter onto the growing central body. Consequently, higher mass densities can be expected for the circumprotostellar disk, as they were computed up to now by models without a barrier (TSCHARNUTER, 1985; MORFILL, 1985; LIN and PAPALOIZON (1985).

This quasiperiodic radial distribution of the orbits of planets in the Solar system and of the satellites in their systems of Jupiter, Saturn and Uranus shall now be described quantitatively to compare these structures with those, as they can be described with equs. (5.157) and (5.159). This comparison of "reality" with theory can be used to fix the yet unknown parameter "n" of the undisturbed radial density profile.

Outer planetary system

It has been shown by MÖHLMANN (1985) that the orbital radii of the outer planets (from Saturn to Pluto) can be described as to be "equally spaced" by $\Delta r = 1.50 \cdot 10^{12}$ m. In other words

$$r_{n+m} = r_n + m \cdot \Delta r . \tag{5.166}$$

if r_n is the orbital radius of the n-th outer planet, and analogously, r_{n+m} that of the $(n + m)$-th outer planet (see Table 1.1). Taking into account now the above mentioned quasiperiodic character of BESSEL's functions (with a periodicity by 2π according to $x \to x + 2\pi$ in $Jp(x)$), it follows by comparing (5.166) and equ. (5.159) with (5.160) that $\beta = 1$ is necessary for the outer planetary system. With $m = 2$ follow then $\alpha = -1, p = 1, n = 5$ and $\varkappa = (M_c/8\pi K H_0^2)^{1/2}$. The resulting solution is then given with

$$f(r) = C_1 r^{-1} J_1(\varkappa r) + C_2 r^{-1} Y_1(\varkappa r), \tag{5.167}$$

where J_1 and Y_1 are Bessel functions of first order and first kind (J_1) and second kind (Y_1). Here, $Jp = K_1 J_p + K_2 Y_p$ has been introduced (see SMIRNOV, 1963). Taking into account the asymptotic periodicity by 2π of these functions, it follows

$$\varkappa r_{n+m} \approx \varkappa r_n + 2\pi m \tag{5.168}$$

leading to

$$\varkappa = \frac{2\pi}{\Delta r}, \tag{5.169}$$

if equs. (5.166) and (5.168) are compared. Numerically, this gives with the "observed" $\Delta r = 1.50 \cdot 10^{12}$ m (see Table 1.1) that $\varkappa = 4.19 \cdot 10^{-12}$ m^{-1}. Using now the "thermal model" (equ. 5.70b), there follows with $H^2 = c_0^2 r^2/\gamma M_c = H_0^2 r^2$ that $H_0^2 = 8.65 \cdot 10^{-3}$.

$$H/r = 9.3 \cdot 10^{-2} \approx 10^{-1}. \tag{5.170}$$

It has to be noted here that comparable results follow for different parameters m and n.

In other words, i.e. the ratio between "thickness" of the disk and the radial distance is of the order of 10^{-1}. This corresponds to an angle of about $\pm 5.3°$, as seen from the Sun. These numerical values give the possibility to derive the density in the outer disk. It follows with $\varkappa^2 = M_c/8\pi K H_0^2$ and $\varkappa = 4.19 \cdot 10^{-12}$ m^{-1} that $K = 2.62 \times 10^{53}$ kg m^2, leading to

$$\varrho(r) = 2.62 \cdot 10^{53} r_{[\mathrm{m}]}^{-5}. \tag{5.171}$$

Numerically follows at pre-Saturnian distances $\varrho \approx 4.4 \cdot 10^{-8}$ kg/m^3 and for pre Neptunian regions $\varrho \approx 1.4 \cdot 10^{-10}$ kg/m^3. The related surface densities $\sigma = 2\sqrt{2}\,\varrho H$ are given then with $\sigma \approx 1.7 \cdot 10^4$ kg/m^2 and $\sigma \approx 1.7 \cdot 10^2$ kg/m^2 respectively. These values are slightly greater than those of WEIDENSCHILLING's "solar composition model" (WEIDENSCHILLING, 1983) in the Saturnian region, but they agree with WEIDENSCHILLING's result for Neptun.

The total mass of the outer disk can be estimated now. It follows with $m_d = 2\pi \int dr r \sigma$ that

$$m_d \approx 2\sqrt{2}\, K H_0^2 r_i^{-2} \approx 2.2 \cdot 10^{29}\ \mathrm{kg}, \tag{5.172}$$

when an inner radius r_i of the outer disk of $r_i = 10^{12}$ m is used

The extrema of $f(r)$, as it is given with equ. (5.167), follow from

$$0 = -C_1 J_2(\varkappa r) - C_2 Y_2(\varkappa r). \tag{5.173}$$

This leads with the known values for $\varkappa$ and r for the outer planets to $C_1 \approx C_2$.

Summarizing, it can be stated that the derived properties of the outer preplanetary disk are compatible with observational data and independently derived results. The essential new result is, that this disk was more massive than it has been assumed in earlier theories. But the resulting mass of the outer disk of about 10 percent of the present Solar mass seems to be an acceptable result, but this is a serious challenge to theories of the formation of protostars and circumprotostellar disks.

Inner planetary system

It has been shown by LITZROTH (1980) und MÖHLMANN (1981) that the inner planetary system and the satellite systems of Jupiter, Saturn and Uranus seem to be ordered radially according to

$$(v_0)_{n+m} = (v_0)_n + m(\Delta\bar{v}), \tag{5.174}$$

where $(v_0)_n = (\gamma M_c/r_n)^{1/2}$ is the orbital velocity of the n-th planet (satellite) — see also Table 1.1. Using now again the quasiperiodic character of Bessel's functions, it follows by comparing equ. (5.159) with (5.174), that $\beta = -\frac{1}{2}$ is necessary to describe these four systems. Consequently, there follows from equ. (5.160) with $m = 2$ that $n = 2$, $\alpha = 1/2$, $p = 1$ and $\varkappa = (M_c/2\pi K H_0^2)^{1/2}$.

Consequently, the radial distribution is given with

$$f(r) = C_1' r^{-1/2} J_1(\varkappa r^{-1/2}) + C_2' r^{-1/2} Y(\varkappa r^{-1/2}). \tag{5.175}$$

The extrema follow then from

$$0 = C_1' J_0(\varkappa r^{-1/2}) + C_2' Y_0(\varkappa r^{-1/2}), \tag{5.176}$$

or, by using the asymptotic expansions of J_0 and Y_0

$$C_1'/C_2' \approx -\tan(\varkappa r^{-1/2} - \pi/4). \tag{5.177}$$

The value of $\varkappa$ can be determined now by comparing (5.174) and

$$\varkappa r_{n+m}^{-1/2} = \varkappa r_n^{-1/2} + 2\pi m, \tag{5.178}$$

what is a consequence of the quasiperiodicity by 2π of Bessel's functions. This comparison gives

$$\varkappa^2 = \frac{4\pi^2 \gamma M_\odot}{(\Delta\bar{v})^2}. \tag{5.179}$$

($M_\odot$ — mass of the present Sun).

With $(\Delta\bar{v}) = 5.27 \cdot 10^3$ m/s follows for the inner planetary system $\varkappa^2 = 1.89 \times 10^{14}$ m. This gives with the known planetary orbital radii from (5.177) that $C_1/C_2 \sim -0.2$ or $|C_2| > |C_1|$. Therefore, and as a first approximation, the solution

$$f(r) = C_2 r^{-1/2} Y_1(\varkappa r^{-1/2}) \tag{5.180}$$

shall be used for the following discussion. It can be shown that this is correct also for the three planetary satellite systems. Numerically follows with $\varkappa^2 = M_c/2\pi K H_0^2$ for the density

$$\varrho = K r^{-2} = 9.74 \cdot 10^{16}/r^2_{[\mathrm{m}]} . \tag{5.181}$$

This corresponds to a density of $2.9 \cdot 10^{-5}$ kg/m³ at pre-Merkurian regions and of $1.6 \cdot 10^{-7}$ kg/m³ at pre-Iovian distances. The surface mass densities are then $\sigma = 4.4 \cdot 10^5$ kg/m² and $3.3 \cdot 10^4$ kg/m² respectively. Also these values are greater up to one order of magnitude than that of the "solar composition model" of WEIDENSCHILLING (1983).

The resulting mass of this inner disk is then given with

$$m_d \approx 4\pi \sqrt{2}\, K H_0 r_a \approx 1.6 \cdot 10^{29}\ \mathrm{kg}, \tag{5.182}$$

when the outer radius of the inner disk is taken as $r_a = 10^{12}$ m. This corresponds to about 10% of the present Solar mass. Consequently, the total disk mass is of about 20% of the present Solar mass. This can be an indication, that slightly more than 50% of the Solar mass were accumulated when the above described planetogonic processes took place. But these estimations are not precisely enough to determine t_p more accurately.

Satellite Systems of Jupiter, Saturn and Uranus

Using the above described formalism, as it was applied to the inner planetary system, there follows with $(\Delta\bar{v}) = 3.04 \cdot 10^3$ m/s for the properties of the Iovian pre-satellite disk (when the "thermal model" $T[K] = 5 \cdot 10^{10}/\mathrm{r}\,[\mathrm{m}]$ and $c^2 = 2.89 \cdot 10^{14}/r$ is used), that $\varkappa^2 = 5.40 \cdot 10^{11}$ m, $H = 6.58 \cdot 10^{-2} r$, $H_0^2 = 4.33 \cdot 10^{-3}$, and

$$\varrho(r) = 6.81 \cdot 10^{16}/r^2 . \tag{5.183a}$$

Here $M_c = 1 \cdot 10^{27}$ kg has been used. The resulting densities are at pre-Io-distances $\varrho \approx 3.8 \cdot 10^{-1}$ kg/m³ and $\varrho \approx 1.9 \cdot 10^{-2}$ in the pre-Callisto region. The related surface densities are then $\sigma \approx 3.0 \cdot 10^7$ kg/m² and $\sigma \approx 6.7 \cdot 10^6$ kg/m² respectively. The total disk mass is then given with

$$(m_d)_{♃} \approx 1.6 \cdot 10^{26}\ \mathrm{kg}, \tag{5.183b}$$

when $r_a = 2 \cdot 10^9$ m is used.

With the same "thermal model" follows with $(\Delta\bar{v}) = 1.47 \cdot 10^3$ m/s for the presatellite disk of Saturn $\varkappa^2 = 6.93 \cdot 10^{11}$ m. Using $H_0^2 = 1.44 \cdot 10^{-2}$ and $M_c = 3 \cdot 10^{26}$ kg follows

$$\varrho(r) = 4.78 \cdot 10^{15} r^{-2}_{[\mathrm{m}]}, \tag{5.184}$$

giving densities of $\varrho \approx 1.4 \cdot 10^{-1}$ kg/m³ at pre-Mimas distances, and $\varrho \approx 3.2 \times 10^{-3}$ kg/m³ in the pre-Titan region. The surface densities are $\sigma \approx 8.8 \cdot 10^0$ kg/m² and $\sigma \approx 1.3 \cdot 10^6$ kg/m² respectively.

The resulting mass of the pre-satellite disk of Saturn is given with

$$(m_d)_{♃} \approx 2.0 \cdot 10^{25}\,\mathrm{kg}, \tag{5.185}$$

when $r_a = 2 \cdot 10^9$ m is used.

For the Uranian pre-satellite system $\varkappa^2 = 2.25 \cdot 10^{11}$ m can be derived, when $\Delta v = 1.01 \cdot 10^3$ m/s is used. The resulting radial density-profile is given then with $M_c = 5 \cdot 10^{25}$ kg and $H_0{}^2 = 4.98 \cdot 10^{-2}$ by

$$\varrho(r) = 7.1 \cdot 10^{14}/\mathrm{r}^2_{[\mathrm{m}]}\,. \tag{5.186}$$

Consequently, in the pre-Miranda region the density was of $\varrho \approx 4.2 \cdot 10^{-2}$ kg/m³, and that at pre-Titania distances of $\varrho \approx 3.7 \cdot 10^{-3}$ kg/m³. The resulting surface mass densities are given then with $\sigma \approx 3.4 \cdot 10^6$ kg/m² and $\sigma \approx 1.0 \times 10^6$ kg/m² respectively. The total disk mass is given with

$$(m_d)_{⛢} \approx 3 \cdot 10^{24}\,\mathrm{kg} \tag{5.187}$$

when $r_a = 1 \cdot 10^9$ m is used.

Ofcourse, these values can not be understood as to be very exact. They give the numerical consequences of the above outlined planetogonic scenario, but with uncertainties due to modelapproximations as $M_c(t_p) \approx 0.5 M_c$ (final) and as the "thermal models". So, these values probabely are only slightly more accurate than their order of magnitude. Nevertheless, the derived results agree well with observable properties of the above mentioned systems. This is a serious argumentation in favour of the above outlined "planetogonic scenario".

5.3.6. Rotation and Azimutal Instability

According to the planetogonic model, described in the foregoing sections, protoplanetary (and planetary) bodies evolve from temporary ringlike configurations which are gravitationally unstable. The "feeding" of the growing planet and the resulting planetary growth from the ring matter should have proceeded in an ordered and non-stochastic way. This can be concluded from regularities especially in planetary rotation.

Parallelism of orbital angular momentum and spin of nearly all the planets and satellites is an even "classical" planetogonically relevant property of these bodies. The planetogonic relevance becomes clear by taking into account, that the local angular velocity $\boldsymbol{\omega}_l = \dfrac{1}{2}\,\mathrm{curl}\,\boldsymbol{v}$ of a local region of a preplanetary disk (corotating at $r = R$ with the local Keplerian angular velocity $\Omega_\mathrm{K} = (\gamma M_c/r^3)^{1/2}$), and which accumulates to a later planet, is given for a corotating observer with $\boldsymbol{\omega}_l = \boldsymbol{e}_z \dfrac{\partial}{\partial x} v_y = -\boldsymbol{e}_z \dfrac{3}{4}\,\Omega_\mathrm{K}$ (see section 5.3.3.), what is antiparallel to its orbital angular momentum. Consequently, the observational fact of parallelism of spin and orbital angular momentum requires, that

— either the spin of all the planets and satellites has been reorientated after their formation in a local region (eventually by tidal action of the central body, as discussed by Farinella and Paolicchi, 1978), what is rather questionable as a general process,

– or, and in accordance with the above described model the growth of planetary bodies from rings in the gravitational field of a central mass occurs "non-isotropically" on selected "capture-orbit-bands" and not stochastic (isotropic), as it is shown schematically in Fig. 5.15. The existence of selected capture-bands has been described by Giuli (1968). They are caused by the influence of the central body on the orbits near to the growing body.

A further remarkable property of planetary rotation has been discussed by Alfvén and Arrhenius (1976), namely the so-called "Spin-Isochronism", what indicates, that the rotational period is of the same order of magnitude (from some hours to about some ten hours) for the most of the bodies in the Solar system (with very different masses between 10^{27} kg for Jupiter and 10^{16} kg for smaller asteroids). Also this property should be connected with the formation of these bodies, since there is probabely no effective later mechanism, which could bring all the rotation periods into the same order of magnitude.

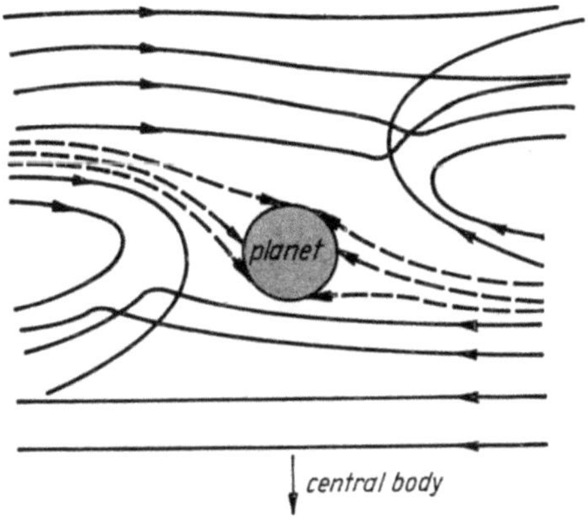

Fig. 5.15. Schematical sketch of orbital pathes for particles, accreting at a growing planet (seen from an observer, corotating with the planets Keplerian motion) and causing prograde rotation by the interplay of the gravitation of the central body and the planet and the centrifugal force (with respect to the central body).

As will be shown now, "Spin-Isochronism" can be understood within the scenario of gravitationally focussed protoplanetary growth from selected bands (cf. Fig. 5.15). The angular momentum gain of the growing planet (of radius R) from the infalling matter of mass $\mathrm{d}m$ is given by $\mathrm{d}L = \beta Rv\,\mathrm{d}m$, where β is a parameter of the order of 10^{-1}, resulting from the average of the (different frequent) collision angles, and where v is the impact velocity Assuming now, that for an estimation of orders of magnitude v can be represented by the velocity $v = (\gamma M_c/R)^{1/2}$ of a captured body, the resulting total angular momentum is $L = \frac{3}{5}\,\beta\gamma^{1/2}(3/4\pi\varrho)^{1/6}\cdot m^{5/3}$, where m is the mass of the planet.

Equating this with the angular momentum of a rotating planet (in the approximation of a spherical body with constant density ϱ) $L_{\mathrm{pl}} = \frac{2}{5} mR^2\omega_{\mathrm{pl}}$, it follows with an estimated $\beta \sim 1/5$ for the angular velocity of rotation

$$\omega_{\mathrm{pl}} = \frac{3}{10}\,\gamma^{1/2}\left(\frac{4\pi\varrho}{3}\right)^{1/2}. \tag{5.188}$$

This relation describes indeed something like "Spin-Isochronism", it depends only via the only weakly variing $\varrho^{1/2}$ on the averaged density of the planet, but not on its mass or diameter. The corresponding numerical values are given in Table 5.12. (cf. Möhlmann, 1983).

Table 5.12. Angular velocities in the planetary system (Mercury, Venus and Pluto/Charon have not been taken into account due to their pecularities)

Planet	ω_{real} (10^{-4} sec^{-1})	ω_{Pl} (10^{-4} sec^{-1})
Earth	0.727	3.75
Mars	0.709	3.16
Ceres	1.92	2.02
Vesta	3.46	2.52
Jupiter	1.78	1.83
Saturn	1.71	1.34
Uranus	1.62	1.82
Neptune	1.16	2.05

It is interesting to note here, that Table 5.12 fits quite well the observed "Spin-Isochronism", if on takes into account the uncertainties in collision angle and collision velocity. In spite of these uncertainties the physical process of gravitationally focussed growth, what was based on the derivation of equ. (5.188) seems to be the cause for the observed isochronism of spins in the planetary system.

It is interesting to note here, that Earth and Mars rotate relatively too slow. For the Earth the present value of ω_{real} could be understood as to be caused by angular momentum transport to the lunar orbit. But what was the case with Mars. Had he lost a greater satellite (as Pallas) by tidal action (Möhlmann, 1983a)?

The above outlined scenario of planetary origin by gravitationally focussed growth is necessarily connected with a heating of the growing protoplanet due dissipation of mechanical gravitational energy, released by the growth (Möhlmann, 1982a) (for more details see section 4.2.3.). This can be estimated by assuming a mass flow $\dot{m} = 4\pi\beta\varrho R^2 v$, where $\beta < 1$ indicates, that matter is accumulated only at a part of the surface in the non-isotropic case (see Fig. 5.15).

Having in mind the uncertainties in initial and boundary conditions, in material properties and in interpretations of observations, it is sufficient for a rough analysis of these processes to take into account only the gravitational energy gain, $\gamma \dot{m} M/R$, the heating due to impacting matter (neglecting latent heat), $\dot{m}\bar{c}T$, and the radiative energy loss $4\pi\sigma R^2 T^4$ (see section 4.2.3.). With $v = (\gamma M/R)^{1/2}$, neglecting an initial relative velocity of the accreting matter, it follows from this energy balance for the mass density of the accreting matter

$$\varrho = \frac{\sigma T^4}{\bar{\beta}\left[\frac{4}{3}\,\pi\gamma R^3\bar{\varrho} - \bar{c}TR\right]\left(\frac{4}{3}\,\pi\gamma\bar{\varrho}\right)^{1/2}} \tag{5.189}$$

where $\bar{\varrho}$ is the mean density of the accreted body of radius R and surface temperature T. $\bar{c}$ represents the heat capacity. Taking into account now the observational fact that the lunar surface was molten, temperatures exceeding 1100 °C should have existed there during the accretion period. The mass density, necessary to generate these temperatures is given according to equ. (5.189) in Table 5.13, where $\bar{\beta} = 0.5$ and $\bar{\varrho} = 3.35 \cdot 10^3$ kg/m³ and $\bar{c} = 1.24$ Ws/kg are used. The resulting ϱ is given in kg/m³ (Möhlmann, 1982). It is a remarkable matter of fact, that these densities are well comparable with those, as they have been derived by the planetogonic scenario for the inner disk (Table 5.9).

Table 5.13. Accretional mass densities for different models of lunar radius and surface temperature

R (10^3 km)	$T = 1200$ °C	$T = 1100$ °C
1.34	$6.0 \cdot 10^{-5}$	$4.5 \cdot 10^{-5}$
1.44	$4.8 \cdot 10^{-5}$	$3.6 \cdot 10^{-5}$
1.54	$3.9 \cdot 10^{-5}$	$2.9 \cdot 10^{-5}$

The duration of this growth towards planetary bodies can be estimated for the above discussed example by

$$\dot{M} = 4\pi R^2 \varrho v \approx \left(\frac{12\pi\gamma}{\bar{\varrho}}\right)^{1/2} \varrho M \tag{5.190}$$

giving as a characteristical growth-time t_g

$$t_g \approx \left(\frac{\bar{\varrho}}{12\pi\gamma}\right)^{1/2} \varrho^{-1}. \tag{5.191}$$

This is of the order of (10^3 – 10^4) years, indicating again a very fast planetary growth. It shall be noted here, that these orders of magnitude for protoplanetary growth are comparable with those, as they have been estimated in section 5.3.3. These independent estimations led to the conclusion of a fast growth of planetary bodies with a characteristical order of magnitude for protoplanetary growth of 10^4-10^5 years, indicating a direct growth of these bodies from gas and dust by gravitationally focussed accretion.

5.3.7. Growth of Planetesimals

The planetogonic scenario, as it has been developed in the foregoing sections, describes the possible growth of planets from an early protostellar disk by gravitational instabilities with giant protoplanets as predecessors of later planetary bodies. Geochemical evidence for the reality of an early and more massive "proto-Earth" has been discussed in part 3. On the other side, planetesimals, as they are assumed to be the essential predecessors of planets within the frame of stochastic (collisional) accretion theories, do not appear in this planetogonic scenario.

This seems to be a contradiction to the observed fact that the surface structure of planets (as Mercury and Luna) and of planetary moons shows clear evidence for impact events. The diameters for the impacting bodies seem to have been ranged up to orders of some 10^4 m -10^5 m. This is interpreted as an essential argument in favour of a collisional planetary growth by impacting planetesimals. But it will be shown in the following, that the reality of these early impact processes is not in contradiction with the above outlined planetogonic scenario.

At first, the results from investigations of the cratering history of Luna shall be mentioned. A surprising result is that the maximum "bombardement" of the lunar surface took place as late as $3.9 \cdot 10^9$ years ago (see part 3) and not directly connected with the formation processes of Luna ($4.5 \cdot 10^9$ years ago). This can be understood as an indication for a later formation of small (atmosphereless) solid bodies from a relatively massless later disk of yet interacting particles, (in the sense of Safronov), following the earlier phase of a relatively mass-rich disk of gas-plasma, dust and particles when the protoplanets were formed. These "second generation" bodies could be responsible then for the impact-caused surface features of planets and satellites.

The physical conditions in this thin disk can be characterized in that way, that the influence of the gaseous contribution, which is essential in an earlier thick disk, should have been decreased. This has to be expected as a consequence of condensation of gases at grains, inward and outward directed outflow of gas from the disk and the later heating and repelling action of protostellar radiation and winds. The resulting thin disk should consist mainly of grains and particles, interacting by gravitation and collisions. Consequently, action of thermal pressure should have become unimportant ($c = 0$). Also friction between particles and gas has been vanished.

The possibilities for gravitational interactions of the particles of this disk can be estimated on the basis of the dispersion-relation for a two-dimensional thin disk $\left(\varrho_0 \sim \sigma_0\, \delta(z)\right)$, as it has been discussed in chapter 5.3.2.

It follows with $k_x^2 = k^2 = k_z^2 = \lambda_c^{-2}$ for the above characterized disk ($c = 0$) from equ. (5.133) that the scale λ_c (which characterizes the radial dimensions of regions where self-gravitation dominates over (disrupting) tidal action of the central body) is given by

$$\lambda_c = \frac{2\pi\gamma\sigma_0}{\Omega_K^2}. \tag{5.193}$$

This relation has been derived and discussed by Goldreich and Ward (1973). It indicates, that also for relatively low densities σ_0 of particle disks (as they have to be assumed for this "relict disk") a gravitationally induced local "accretional" growth can have been possible. This was the main idea of the Goldreich-Ward formation-mechanism of planetesimals.

But, and as has been shown by Goldreich and Ward (1973), such a differentially rotating region is not able to collapse unimpeded as a whole. This is due to conservation of (local) angular momentum of the differentially rotating region. According to Goldreich and Ward (1973), only small fragments of this region could be able to collapse directly. The characteristical dimensions of these fragments are approximately of some 10^4 m, giving solid bodies ("first generation" planetesimals) with $m < 2 \cdot 10^{11}$ kg and diameters of some 10^2 m (Goldreich, Ward, 1973). According to these authors, "second generation" planetesimals shall be formed then from gravitationally bound clusters of first generation planetesimals. The diameters of these secondary bodies with masses of some 10^{13} kg should be of the order of some kilometers.

The following discussion gives only slightly different results for the rotationally influenced gravitationally caused growth of planetesimals from dust and grains. It shall be assumed as a model, that a fragment with radius r_a of the gravitationally bound region of diameter λ_c collapses. Its original mass density is that of the undisturbed disk. The (retrograde) local angular velocity of this fragment with original Keplerian motion is of the order of $\Omega_K = (\gamma M_c/R^3)^{1/2}$. This fragment collapses towards an equilibrium state where rotation compensates the gravitation of the fragment. Internal (thermal) motions of the fragment shall be neglected. Angular velocity and characteristical radius of this equilibrium shall be characterized by ω_e and r_e respectively. Angular momentum conservation leads with $\Omega_K{}^2 r_a{}^2 = (\gamma\pi r_a{}^2\sigma_0 r_e)^{1/2}$ then to $r_a{}^2/r_e = \pi\sigma_0 R^3/M_c$, describing a rotating fragment or subcloud which probably can not reach the final state of a solid body. This final state seems to be possible only, if this cloud was able to collapse to a density of the order of $\varrho_s \approx 10^3$ kg/m^3, typical for solid matter. With the corresponding surfaces density $\sigma_s = \frac{4}{3} r_s\varrho_s$ (from the estimation $\pi\sigma_s r_s^2 = \frac{4}{3}\pi_s r_s^3\varrho_s$) of the solid body of radius r_s follows for its mass $m = r_a{}^2\sigma_s$ and $r_a{}^2 = 4r_s{}^3\varrho/3\sigma_0$, giving with $r_e = r_s$ as the condition for the formation of solid bodies the relation $r_s = \sigma_0(3\pi R^3/4M\varrho_s)^{1/2}$ for the maximum radius of "first generation" planetesimals. This is of the order of $r_s \sim 10^3$ m for the planetary system. The growth of these planetesimals should have happened slowly and gently. This seems to be in agreement with properties of carbonaceous chondrites which should have been formed "gently" by agglomeration instead of violent collisional accretion. In agreement with the results of Goldreich and Ward (1973) the mass of these gently accumulated planetesimals is of the order of 10^{13} kg. Their composition could have been similar to that of carbonaceous chondrites. Probably they interact with those originally greater fragments, which reached an equilibrium at $r_e > r_s$ and failed to evolve towards solid bodies. Consequently only a part of the fragments should form solid bodies. Another part should be dispersed then by interaction with these first generation planetesimals, which themselves could grow by this interaction, forming slightly modified "second generation" planetesimals. The km-sized planetesimals have a Hill-sphere of the order of $10^5 - 10^6$ m, what is of the order of the average distance ($\sim r_a$) of the planetesimals. Consequently, these planetesimals are yet interacting collectively by their gravitation. This should lead to the formation of the above mentioned "second generation" planetesimals with diameters of the order of about 10 km. A further stochastic growth is possible then by collisional accretion as it has been ontlined by SAFRONOV (1969). The resulting bodies (and earlier planetesimals) can have led then to the above mentioned impact-features, observable on planetary surfaces.

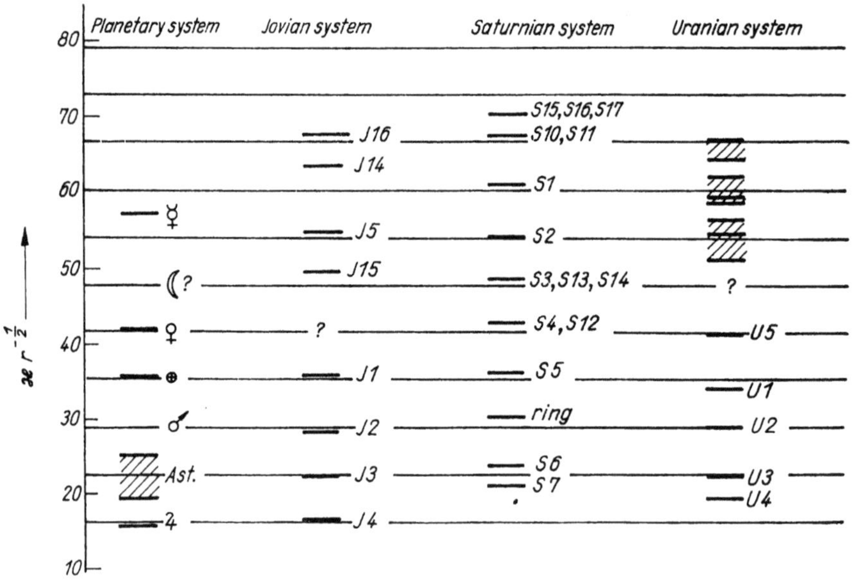

Fig. 5.14. Theoretically determined orbital levels (thin lines) and real orbital radii (short thick lines) of the satellites in the inner planetary system and the systems of Jupiter, Saturn and Uranus. The gap in the Iovian system might be caused by disturbances due to Io (J 1) (which is locked now into resonance with Europa (J 2). The reality of gap in the Uranian system is a yet open question. Up to now, Voyager data were not investigated in this orbital region for yet unknown satellites (SMITH, 1986). At the position, indicated by "ring" in the Saturnian system, Mrs. Vasundhara (1984) has found a ring by Earth-bound optical (occultation) observations.

Table 5.9. Model parameters for the four forming system (proto-Sun, proto-Jupiter, proto-Saturn, proto-Uranus) of central body and disk when gravitational structuring can have been effective. The parameters for the inner Solar system are taken from results, derived in section 5.3.4. The other data follow from section 5.3.6.

	$0 = (M_0)_0/$ $6r^3$ (10^{-5} kg m^{-3})	$H = (4c^2r^3)/$ $(M_c)_0)^{1/2}$ (10^{10} m)	$\sigma_0 = \sqrt{\pi}\, H\varrho_0$ (10^5 kg/m^2)	Parameters
Mercury	34.80	1.06	65.4	$(M_c)_0 = 1.28 \cdot 10^{30}$ kg
(Luna)	12.32	1.63	35.6	
Venus	5.39	2.30	22.0	
Earth	2.03	3.45	12.4	$(m_{disk})_i = 0.33 \cdot 10^{30}$ kg
Mars	0.57	5.85	5.91	$c^2 = 2.94 \cdot 10^{12} r^{-1/2}$ (m)
Asteroids	0.14	10.5	2.61	$r_a = 10^{12}$ m
Jupiter	0.014	27.1	0.67	$r_i = 5 \cdot 10^{10}$ m
Saturn	0.004000	36.6	0.260	$M_c = 1.61 \cdot 10^{30}$ kg
Uranus	0.000174	104	0.032	
Neptune	0.000023	204	0.0083	$T = 2 \cdot 10^2$ K
Pluto	0.000007	311	0.0036	$c^2 = 1.23 \cdot 10^6$ m^2/s^2
	(10^{-2} kg/m^3)	(10^8 m)	(10^8 kg/m^2)	$(M_c)_0 = 1.42 \cdot 10^{27}$ kg
Io	100	1.40	2.48	$m_{disk} = 0.48 \cdot 10^{27}$ kg
Europa	24.9	2.50	1.10	$c^2 = 1.27 \cdot 10^{11} r^{-1/2}$ (m)
Ganymed	6.13	4.48	0.49	$r_i = 1.1 \cdot 10^8$ m
Callisto	1.13	9.07	0.18	$r_a = 2 \cdot 10^9$ m
	(10^{-2} kg/m^3)	(10^8 m)	10^8 kg/m^2)	
Mimas	245	1.10	4.78	$(M_c)_0 = 2.97 \cdot 10^{26}$ kg
Enceladus	117	1.50	3.11	$m_{disk} = 2.73 \cdot 10^{26}$ kg
Tethys	61.4	1.96	2.13	$c^2 = 1.27 \cdot 10^{11} r^{-1/2}$ (m)
Diane	29.4	2.66	1.39	$r_i = 1.4 \cdot 10^8$ m
Rhea	10.8	4.04	0.77	$r_a = 3.6 \cdot 10^9$ m
Titan	0.86	11.6	0.18	
Hyperion	0.49	14.7	0.13	
	(10^{-2} kg/m^3)	(10^8 m)	(10^8 kg/m^2)	
Miranda	125	0.84	1.86	$(M_c)_0 = 5.19 \cdot 10^{25}$ kg
Ariel	38	1.37	0.92	$m_{disk} = 3.61 \cdot 10^{25}$ kg
Umbriel	14	2.07	0.51	$c^2 = 3.18 \cdot 10^{10} r^{-1/2}$ (m)
Titania	3	3.84	0.2	$r_i = 0.5 \cdot 10^8$ m
Oberon	1	5.52	0.1	

Table 5.10. Orbital radii for outer planets

Planet	r_{real} (10^{11} m)	r_{model} (10^{11} m)
Saturn	(14.33)	(14.33)
Uranus	28.84	29.44
Neptune	45.09	44.55
Pluto	59.66	59.66
?		74.77
?		89.88
?		105
?		120

$L = (r_n - r_{n-1})/2\pi$ has been determined from Saturn, Uranus and Neptune

Table 5.11. Theoretical and real orbital radii of the satellites of the Sun, Jupiter, Saturn and Uranus (s = number of the zeros of $J_1(\varkappa r^{-1/2})$, $(\varkappa r^{-1/2})_s$ is the argument at the zero-value of J_1)

s	$(r/\varkappa^2)_s$	Planetary system			Jovian system			Saturnian system			Uranian system		
		Ident.	r_{real} (10^{11} m)	r_{model} (10^{11} m)	Ident.	r_{real} (10^8 m)	r_{model} (10^8 m)	Ident.	r_{real} (10^8 m)	r_{model} (10^8 m)	Ident.	r_{real} (10^8 m)	r_{model} (10^8 m)
1	0.0681106			(129)			(367)			(600)			(156)
2	0.0203176												
3	0.0096619			(18)			(52)			(85)			(22)
4	0.0056332												
5	0.0036862	♃	7.78	6.69	J4	18.8	19.88	S8	35.61	32.49	U4	5.86	8.45
6	0.0025989												

7	0.0019304	Aster	3–5	3.64	J3	10.7	10.41	S7	14.81	17.01	U3	4.38	4.43
8	0.0014903												
9	0.0011852	♂	2.28	2.24	J2	6.71	6.39	S6	12.22	10.45	U2	2.67	2.72
10	0.0009651												
11	0.0008010	⊕	1.50	1.51	J1	4.22	4.32	c)	7.5	7.06	U1	1.92	1.84
12	0.0006755												
13	0.0005774	♀	1.08	1.09		b)	3.11	S5	5.27	5.09	U5	1.30	1.32
14	0.0004992												
15	0.0004358	?	a)	0.82	J15	2.22	2.35	S4, 12	3.77	3.84	?	—	1.00
16	0.0003838												
17	0.0003406	☿	0.579	0.64	J5	1.81	1.84	S3, 14 S14	2.95	3.00	e)	e)	0.78
18	0.0003043												
19	0.0002735		average of $\varkappa^2$		J14	1.34	1.47	S2	2.38	2.41	f)	f)	0.63
20	0.0002471		from ♀, ⊕ and ♂										
21	0.0002245		$\varkappa^2$ — 1.887		J16	1.18	1.21	S1	1.86	1.98	ε-ring g)	0.51	0.51
22	0.0002047		$\times 10^{14}$ m										
23	0.0001876				average of $\varkappa^2$			S10, 11	1.51	1.65			
24	0.0001723				from J1, J2, J3, J4,						average of $\varkappa^2$		
25	0.0001590				$\varkappa^2 = 5.39 \cdot 10^{11}$ m			S15, 16 S17	1.40	1.40	from U1, U2, U3 and U5		
								average of $\varkappa^2$ from S3, S4 and S5 $\varkappa^2 = 8.81 \cdot 10^{11}$ m			$\varkappa^2 = 2.29 \cdot 10^{11}$ m		

a) possible orbit of the early Moon (Luna)
b) possibly disturbed by Io (only 120000 km distance)
c) position coincides nearly with the newly found outer Saturnian ring (see VASUNDHARA et al. 1985)
d) possibly disturbed by Titania = U3
e) 1985 U1 – 0.8598; 1986 U4 – 0.6992; 1986 U5 – 0.751
f) 1986 U1 – 0.6609; 1986 U2 – 0.6435; 1986 U3 – 0.6175; 1986 U6 – 0.627; 1986 U9 – 0.591
g) ε-ring at 0.51; 1986 U7 – 0.493; 1986 U8 – 0.533

(e–g: see Table 1.4)

6. The Planetogonic Scenario (Summary)

The results of the foregoing chapters shall be summarized now within the following "planetogonic scenario", what shall be applicable to the description of origin and early evolution as well of the planetary system as (in an analogous way) of the satellite systems of Jupiter, Saturn and Uranus. The following planetogonic process-sequence is postulated:

1. Gravitational collapse of a rotating (magnetized) fragment of an interstellar cloud of gas, dust and plasma.

The description of these collapse processes is not in the center of interest of this book. Therefore, only the basic elements of these processes have been described in chapter 5.1. and its sections. It shall be noted here, that these processes, and especially the influences of viscosity, rotation and magnetic fields under realistic 3-dimensional conditions are not yet finally describable in detail. Nevertheless and in spite of unresolved details, there is no doubt that star formation via gravitational collapse of a fragment of an interstellar cloud is a real astrophysical process. It's reality is well founded by as well astronomical observations as theoretical models of star formation. The final state of this collapse, namely the protostar and its environment give first conditions and contraints which have to be taken into account in planetogonic models.

2. The formation of a protostar (and of a protoplanet) as the "central body" of a system is connected with the evolution of internal convection within the protostar and with resulting, possibly time-dependent, large-scale circumstellar (circumprotoplanetary) magnetic fields with typical magnetic inductions ot the order of 10^2 G $= 10^{-2}$ Vs m^{-2} at distances of some protostellar radii (see especially section 5.1.6.). The environment of the protostar has to be characterized furthermore by a circumprotostellar shell of hot plasma ($10^3\,K$ to $10^4\,K$) with electron densities of the order of 10^{16} m^{-3} and dimension of some protostellar radii (10^9 m-10^{10} m). The radii of protostellar shells increase while the protostar evolves. This seems to be connected with an increase in the rotation period of the protostar (outward transport of angular momentum). These observational facts can be explained, if the outer boundary of the circumprotostellar disk is identified with the Alfvén-surface of partial corotation (see section 5.2.2.). The analogue modeling of protoplanetary growth is a yet open problem.

Outside of the circumprotostellar shell cooler matter dominates with temperatures of the order of $10^2\,K$ or $10^3\,K$. It consists mainly of gas and dust. The dust may be "primary dust" from the interstellar cloud and "secondary dust" resulting from condensation (followed by mineralization) of matter in the circumprotostellar envirónment.

3. Due to rotation, formation of the protostar (and protoplanet) and its environment should proceed from a disk-like configuration (see chapter 5.2. and its sections, especially 5.2.4. and 5.2.6.). The existence of such more or less equatorial circumprotostellar disks ($H/r \sim 10^{-1}$) has been confirmed by astronomical observations, especially at infrared- and submillimetre wavelengths (see Fig. 1.1 and section 5.1.6.). These disks are enriched relatively by dust. The optical appearance is that of a biconical nebula (see Fig. 5.9). This is due to the fact, that there is an rotationally and magnetically caused outflow and thinning of matter over the poles. Consequently, the regions above the poles are optically thin and the light may leave the system mainly over the poles while the equatorial regions of the disk (including the protostar) remain obscured by the opaque thick disk.

4. The existence of a relatively massive early circumprotostellar thick preplanetary disk, which is yet feeding the protostar, is connected with condensation and partial reheating and resulting complex chemical and mineralogic process sequences (see part 3). The chemical composition of this preplanetary disk is essentially influenced by active phenomena at the central body, as "superflares" (see section 3.4.5.), by the outward flux of "secondary dust" from condensed matter, originating from inside of the protostellar shell (see chapter 1.3. and part 3), by the heating and resulting (partial) evaporation due to electromagnetic radiation and protostellar "T-Tauri-wind" (with resulting sputtering), by the electromagnetic radiation pressure acting on the μm- and sub-μm-particles of the dust (see part 3 and sections in chapter 5.2.) and by transport processes in the disk (eventually under the influence of turbulence).

5. This relatively massive circumprotostellar disk around the yet forming protostar (with m_{disk} of about some tenth of $m_{\text{protostar}}$) can give the first appropriated conditions for radially structuring processes, driven by the self-gravitation of the disk. As a result, the disk should be able to decay into separated (temporary) rings or bands (as a transient state). These rings should become unstable (at as short time-scales as about 10 orbital periods), decaying by the azimutal instability into one (or more) very massive "protoplanets" (see chapter 5.3.).

These forming protoplanets are imbedded in circumprotoplanetary disks which can be formation sites for the satellite systems of these later planets.

Modelling of the evolution towards relatively mass rich disks around yet growing central bodies is a challenge for future work on star formation.

6. An essential process within the frame of this "planetogonic scenario" is the fast growth of protoplanetary bodies from the temporary (or transient) state of rings (or bands). These rings should be gravitationally unstable against disturbances. Their "decay" into growing protoplanets should proceed within time-scales of about ten Keplerian periods. Insofar, this "azimutal instability" is very effective.

7. There is some geochemical evidence that the present Earth resulted from a predecessing hydrogen-dominated proto-Earth (see chapter 3). This hypo-

thesis is favoured by arguments, coming from estimations on the chemistry in hydrogen-dominated media and from the fact, that real masses of elements in the Earth's body indicate the early existence of a more massive proto-Earth, if cosmic abundances are assumed for the original state of the "planet-forming" circumprotostellar disk. This result supports the generalizing conclusion, that quite generally, proto-planets were predecessors of present planets. This gives new aspects for understanding the evolution of the internal structure of planets.

8. The quantitative description of the above outlined planetogonic process-sequences (see chapter 5.3.) leads to the result, that there are two different types of satellite systems in the Solar system. There are at first the satellite systems of Jupiter, Saturn and Uranus and the "inner planetary system" between Mercury and Jupiter, which are radially ordered by a spacing with a constant velocity difference between the orbital radii. On the other side, the outer planetary system seems to be ordered by "equidistant" spacing (see chapter 1.4.).

The mathematical modelling of the gravitationally caused structuring processes leads to the result, that a relatively mass-rich disk with a mass of some tenth of the central bodies mass must have been existent. This means, that these structuring processes, must have taken place relatively early, when the central body was yet forming. At this time the disk was also vertically extended, with a scale-height $H \sim 10^{-1}$ r. For quantitative modeling a radial temperature profile $T \sim r^{-1}$ seems to be a good approximation. Based on these models, there follows by comparing theoretically derived and observationally determined radial structures in the above mentioned systems, that the radial orbital distribution in the outer planetary system can be explained best by using a strong radial decrease of the mass density of the outer disk ($\varrho \sim r^{-5}$), while the radial profile in the four other systems (inner planetary, Iovian, Saturnian, Uranian) was of $\varrho \sim r^{-2}$. The resulting profiles for the surface mass densities in the planetary system are well compatible with the "solar composition"-model of WEIDENSCHILLING (1977), but the absolute values can be greater up to about one order of magnitude than in WEIDENSCHILLING's model.

9. The above mentioned theoretically determined radial structures agree quite well with present structures in the Solar system. It is interesting to note here, that the three orbital levels between the ε-ring and the satellite Miranda in the Uranian system (which were predicted on the basis of this planetogonic scenario in 1984 (MÖHLMANN, 1984) have been verified by Voyager-observations in 1986.

A further interesting result is the existence of a "nonoccupied" orbital level in the inner planetary system (between Mercury and Venus). This could be the original formation site of the Earth's Moon "Luna". This speculation should be investigated more in detail in future work.

It should be mentioned here too, that an originally "empty" level in the Saturnian system has been "filled" meanwhile by the observation of a ring near to the position of this level (VASUNDHARA, 1985).

10. Most of the matter of the "planet-forming" circumprotostellar (circum-

protoplanetary) disk should have fallen down onto the growing central body. Due to theit relatively great masses, the protoplanets were not influenced essentially in their later orbital motion by this process (see part 5.3.).

11. Furthermore, following this early preplanetary state, there is the possibility for the evolution towards a later disk of solid particles after the formation of protoplanets and the outflow of the gaseous matter from the originally more massive disk. Gravitationally caused growth processes up to planetary dimensions are not possible in this late and thin disk. This is mainly due to the low mass density of this disk. But km-sized particles are able to evolve due to local gravitational instability (Goldreich-Ward-instability; see part 5.3.). These "second generation" bodies ("planetesimals") could be the reason for the later bombardment of the planets, leading to yet presently observable impact structures on planetary surfaces. Furthermore, this process can be the reason for the existence of some smaller satellites.

Very probabely, comets are examples of these "second generation" solid bodies from the outer Solar system. Primitive carbonaceous chondritic meteorites may be the corresponding counterparts from the inner Solar system.

12. In connection with the formation of larger bodies (around the Sun or the planets respectively), together with the radius r_A of the circumprotostellar (or circumprotoplanetary) shell (as the inner radius for the disk) the radiusr_R of the "Roche limit" is fundamental for the furthere volution of a disk. The Roche limit describes that charakteristical radius inside of which any greater or planetarysized body would be disrupted by the tidal force of the central body. Consequently, for $r < r_R$ also no accretion could be effective to built up greater bodies, and for all cases with $r_A \leqq r \leqq r_R$ the disk should remain in a state of rings similar to that of Saturn. It is interesting to note, that the above used condition $r_A < r_R$ is fulfilled in the systems of Saturn and Uranus only. This might be an explanation why only these two planets have "primary" rings. The ring of Jupiter result from a decaying satellite, it is of "secondary" type. Furthermore, it follows that for the Sun and the other planets holds $r_A > r_R$. Consequently, no preplanetary matter was deposited inside of r_R and no rings have been evolved. This can explain, why the intensive search for rings of Neptune and the Sun was not succesfull (see section 5.2.3.).

13. The evolution of the originally gaseous and dusty thick preplanetary matter of the disk towards grains, embedded in the gas-plasma envelope of decreasing density, has to be described within the frame of a complex sequence of complicated and partially yet unknown chemical and mineralogical processes (see part 3). Here it should be noted, that a radial gradient in the chemical composition of the disk must evolve due to temperature gradients and radial transport of dust. The gas content of the disk decreases if there are sufficient protostellar heating and repulsion due to the growth of the central body and related radiation pressure (and stellar winds). Only in outer regions of the disk, where also that matter is deposited, which was repelled from inner regions, volatiles and other gases (which were not able to escape by their thermal motion) are able to survive longer. Probabely, a part of them may condense out there on the surfaces of dust particles. These coats of frozen volatiles and other frozen gases might have been existent also in the inner

disk, but they should have been sublimated there, similar to small comets. Consequently, there should appear a tendency for the content of volatiles to be relatively greater in the outer disk. The chemical composition of the grains of the inner disk should be influenced mainly by the possibilities for condensation at these higher temperatures and by transport processes.

14. The solid preplanetary material was appearently a very complex mixture of matter formed by many different synthesis processes. Taking into condensideration only materials as silicates, oxides or metals (not H, He, etc.), the presently known preplanetary matter evolved from supernova condensates, from "indigneous" Solar system condensates (resulting in connection the evolution of the Sun), and from condensates formed from revaporized supernova condensates (see section 3.2.1.). Furthermore, some isotopic anomalies (especially Ne—E) indicate a very short time interval between the nucleosynthetic events and the accretion of the material forming meteorites, above all carbonaceous chondrites (see section 3.4.4.). But this is yet in discussión. the nature of the contact zone between chondrules and matrix material in some carbonaceous chondrites shows, that high temperatures did not exist during their agglomeration. This supports suggestions of a "gentle growth". On the other hand, results from investigations of Ca—Al-rich inclusions (concentration of elements in the surface and innermost parts of chondrules) of carbonaceous chondrites point to violent processes and to partial evaporation and recondensation of solids. These processes should have occured earlier then the investigated meteorites originated (section 3.4.6.).

References

AHRENS, T. J., 1979, J. Geophys. Res. B **84**, 985.

ALFVÉN, H., 1954, On the Origin of the Solar System Oxford University Press, Oxford.

ALFVÉN, H., ARRHENIUS, G., 1976, Evolution of the Solar System, NASA-SP-345, Washington D.C.

ALFVÉN, H., 1981, Cosmic Plasma, D. Reidel Publ. Co (Astroph. and Space Science Library Vol. 82).

ALLEGRE, C. J., 1982, in: Formation of Planetary Systems, ed. A. BRAHIC, Cepadues — Ed. Toulouse, France, 283.

ALLEGRE, C. J., BIRCK, J. L., FOURADE, S., SEMET, M., 1975, Science, **187**, 436.

AMBARTSUMIAN, V. A., 1947, Stellar Evolution and Astrophysics, Acad. Sci. Armenian S. S. R., Erevan.

ANDERS, E., 1972, in: H. REEVES (Ed.) On the origin of the Solar System, Centre National de la Recherche Scientifique, Paris, pp. 179—201.

ANDERS, E., 1981, Proc. R. Soc. London, A **374**, 207.

ANDERSON, J. D., NULL, G. W., BILLER, E. D., WONG, S. K., HUBBARD, W. B., MAC FARLANE, J. J., 1980, Science **207**, 449.

APPENZELLER, I. et al., 1980, Astron. Astrophys., **90**, 184.

ARKANI-HAMED, J., TOKSÖZ, M., 1984, Thermal evolution of Menus. Earth Planet. Inter., **34**, 232—250.

ARTJUSHKOV, E. V., 1970, Phys. Earth Planet. Inter. **2**, 318.

ARVIDSON, R. E., GOETTEL, K. A., HOHENBERG, C. M., 1980, Rev. Geophys. Space Phys. **18**, 565.

AUDOUZE, J., 1983, A few remarks regarding the major isotope anomalies in meteorites, Pre-publication Nr. 45, Institut d'Astrophysique de Paris.

BARDEEN, J. M., 1971, Astrophys. J. **157**, 425.

BAUMGARDNER, J. R., ANDERSON, O. L., 1980, Using the thermal pressure to compute the physical properties of terrestrial planets, preprint, Inst. Geophys. Planet. Phys. Univ. Cal. Los Angeles publication 2056.

BEATTY, J. K., 1981, Sky and Telescope. **62**, No. 5, 430—444.

BEGEMANN, F., 1980, Rep. Prog. Phys., **43**, 1309.

BELTON, M. J., 1980, Icarus **41**, 71.

BERTOUT, 1982, in: Proceedings of the 3rd European IUE Conference, ESA-SP 176.

BETTWIESER, E., 1978, Umschau, **78**, 659.

BIBRING, J. P., LANGEVIN, Y., 1982, in: Formation of planetary systems, A. BRAHIC ed., Paris.

BIRCH, F., 1965, J. Geophys. Res. **70**, 6217.

BLACK, D. C., 1972, Geochim. Cosmochim. Acta, **36**, 377.

BLOCH, D., 1978, in: H. KELM (Ed.), High pressure chemistry, D. Reidel Publishing Company, Dordrecht, 199.
BOCHSLER, P., 1983, Europhysics News, **14**, 1.
BODENHEIMER, P., 1978, Astrophys. J., **224**, 488.
BODENHEIMER, P., 1979, Astron. Astrophys., **74**, 288.
BODENHEIMER, P., 1980, in: IAU Symp. 93, Fund. Probl. in the Theory of Stellar Evol., ed. D. SUGIMOTO, D. Q. LAMB, D. N. SCHRAMM, Reidel Publ.
BODENHEIMER, P., BLACK, D. C., 1979, in: Protostars and Planets (ed. T. GEHRELS) (Tuscon, Univ. of Arizona press), 288.
BOSS, A. P., 1980, Astrophys. J. **237**, 563.
BOSS, A. P., 1982, Astrophys. Journal, **259**, 159.
BOSS, A. P., 1985, Scientific American, V. 252 (1), pp. 40—45.
BRAGINSKI, S. J., 1963, Dokl. Akad. Nauk SSSR **149**, 8.
BRAHIC, A., 1982, ed., Formation of Planetary Systemes, CEPADUES editions, Paris.
BRASS, G. W., HARRISON, C. G. A., 1982, Icarus **49**, 86.
BRECHER, A., 1971, in: The Evolutionary and Physical Problems of Meteorites, NASA-SP-319.
BRECHER, A., 1971, On the primordial condensation and accretion environment and the remanent magnetization of meteorites, NASA-SP-319, Washington D.C.
BRECHER, A., 1972, in: On the Origin of the Solar System, ed. H. REEVES, CNES, Paris; and PH. D. THESIS, Univ. Calif., San Diego.
BRECHER, A., ARRHENIUS, G., 1974, J. Geophys. Res., **79**, 2081.
BRETT, R., 1976, Rev. Geophys. Space Phys. **14**, 375.
BUFFON, G. L. L., 1745, De la formation des planetes, Paris.
BUKOWINSKI, M. S. T., 1976, Phys. Earth Planet. Inter. **13**, 57.
BURY, J. S., 1979, J. Brit. Interplanet. Soc. **32**, 123.
BUSSE, F. H., 1975, Geophys. J. R. astr. Soc. **42**, 437.
BUSSE, F. H., 1976, Phys. Earth Planet. Inter. **12**, 350.
CAFFEE, M. W., GOSWAMI, J. N., HOHENBERG, C. M., and SWINDLE, T. D., 1983, J. of Geophys. Res. **88** Supplement, 267.
CAILLAULT, J.-P., and HELFAND, D. J., 1985, The Astrophys. J. **289**, 279.
CAMERON, A. G. W., 1968, in: AHRENS (ed.) Origin and distribution of elements, Pergamon Press, pp. 125—143.
CAMERON, A., 1973, Accumulation processes in the primitive solar nebula, Icarus, **18**, 407—450.
CAMERON, A., 1973, Icarus, **18**, 407—450, see also: CAMERON (1985).
CAMERON, A. G. W., 1978, Moon and Planets **18**, 5.
CAMERON, A. G. W., 1984, Icarus, **60**, 416.
CAMERON, A. G. W., 1985, in: Protostars and Planets II, eds. D. C. BLACK and M. S. MATTHEWS, University of Arizona Press, 1985, pp. 1073—1101.
CAMERON, A. G., 1977, Icarus, **30**, 447. TRURAN J. W.,
CAP, F., 1970, Einführung in die Plasmaphysik, Bd. 1. Akademie Verlag, Berlin.
CASSEN, P., MOOSMAN, A., 1981, Icarus, **48**, 353.
CHRISTOPHE MICHEL-LEVY, M., 1981, Earth Planet. Sci. Lett., **54**, 67.
CHRISTOPHE MICHEL-LEVY, M., KURAT, G., BRANDSTÄTTER, F., 1982, Earth Planet. Sci. Lett., **61**, 13.

Cisowski, S., Collinson, D., Runcorn, S., Stephenson, A., Fuller, M., 1983, J. Geophys. Res., **88**, A691–A704.

Clayton, D. D., 1977, Icarus, **32**, 255.

Clayton, D. D., 1979, Space Science Reviews **24**, 147.

Clayton, D. D., 1985, Nature, **315**, 633.

Clayton, D. D., Dweck, E., Woosley, S. E., 1977, The Astrophys. J. **214**, 300.

Clayton, R. N., Grossman, L., Mayeda, T. K., 1973, Science, **182**, 485.

Clayton, R. N., Mayeda, T. K., 1977, Geophys. Res. Lett., **4**, 295.

Clayton, R. N., Onuma, N., Grossman, L., Mayeda, T. K., 1976, Earth Planet. Sci. Lett., **30**, 10.

Clayton, R. N., Onuma, N., Ikeda, Y., Mayeda, T. K., Hutcheon, I. D., Olsen, E. J., Molini-Velsko, C., 1983, in: Chondrules and their Origines, ed. E. A. King, Lunar and Planetary Institute, Houston, 37.

Mc Cammon, C. A., Ringwood, A. E., Jackson, I., 1983, Geophys. J. R. astr. Soc. **72**, 577.

Cohen, M., 1973, Mon. Not. Roy. Astr. Soc., **164**, 395.

Cohen, M., Kuhi, L. V., 1979, Astrophys. J. Suppl., **41**, 743.

Cole, G. H. A., 1981, Contemp. Phys. **22**, 397.

Cook, A. H., 1972, Proc. R. Soc. Lond. A **328**, 301.

Coradini, A., Federico, C., Lanciano, P., 1983, Phys. Earth Planet. Inter. **31**, 145.

Cox, L. P., Lewis, J. S., 1980, Icarus **44**, 706–721.

Crozaz et al., 1972, Lunar Science III, 164.

Davies, G. F., 1980, J. Geophys. Res. B **85**, 7108.

Day, K. L. and B. Donn, 1978, Astrophys. J. Lett., **222**, L45.

Dearborn, D., Blake, J., 1985, The Astroph. J. Lett., **288**, L21.

Descartes, R., 1644, Principia Philosophiae, Amsterdam.

Donn, B., 1984, Ber. Bunsenges. Phys. Chem. **88**, 306.

Dorfi, E., 1981, 3D models for self-gravitating rotating magnetic interstellar clouds, MPI-PAE/Astro 288 preprint.

Dziewonski, A. M., Anderson, D. L., 1984, Am. Sci. **72**, 483.

Dziewonski, A. M., Hales, A. L., Lapwood, E. R., 1975, Phys. Earth Planet. Inter. **10**, 12.

Ebeling, W., Richert, W., 1985, phys. stat. sol. (b) **128**, 467.

Eberhardt, P., Jungk, M. H. A., Meier, F. O., Niederer, F. R., 1981, Geochim. Cosmochim. Acta, **45**, 1515.

Eiora et al., 1979, Astron. Astrophys., **74**, 89.

Elsasser, W. M., 1963, in: J. Geiss and E. Goldberg (Eds.) Earth sciences and meteorites, North Holland, Amsterdam, **1**.

Esposito, L. W., O'Cauaghan, M., 1983, Icarus **56**, 439.

Evans, N. J. II, 1981, in: Infrared Astronomy, (IAU Symp. 96, ed. C. G. Wynn-Williams and D. P. Cruikshank, D. Reidel Publ.).

Evensen, N. M., Carter, S. R., Hamilton, P. J., O'Nions, R. K., 1978, Earth Planet. Sci. Lett., **42**, 223.

Farinella, P., Paolicchi, P., 1978, Moon and Planets, **18**, 195.

Fearn, D. R., Loper, D. E., 1981, Nature **289**, 393.

Feigelson, E. D., 1982, Icarus **51**, 155.

Feigelson, E. D., 1982, Icarus, **51**, 155.

FEIGELSON, E. D., DE CAMPLI, W. M., 1981, The Astrophys. J. **243**, L89.
FINK, U., IP, W., 1983, Physik in unserer Zeit **14**, 170.
FINN, G. D., SIMON, T., 1977, Astrophys. J., **212**, 472.
FLASAR, F. M., BIRCH, F., 1973, J. Geophys. Res. **78**, 6101.
FOWLER, W. A., 1985, Science, 226, 922.
FOWLER, W. A., GREENSTEIN, J. L., HOYLE, F., 1962, Geophys. J. **6**, 148.
FRANCK, S., 1980, Ann. Phys. (Leipzig) **37**, 349.
FRANCK, S., 1981a, Adv. Space Res. **1**, 203.
FRANCK, S., 1981b, Moon and Planets **25**, 131.
FRANCK, S., 1982a, Gerlands Beitr. Geophys. **91**, 369.
FRANCK, S., 1982b, Phys. Earth Planet. Inter. **27**, 249.
FRENKEL, J. I., 1957a, Kinetische Theorie der Flüssigkeiten, Deutscher Verlag d. Wissenschaften, Berlin, 510 pp. Engl. ed. Clarendon Press, Oxford, 1946.
FRENKEL, J. I., 1957b, Statistische Physik, Akademie-Verlag, Berlin, 783 pp.
FUKAI, Y., AKIMOTO, S., 1983, Proc. Jpn. Acad. **59**, Ser. B, 158.
GAHM, G. F., 1980, in: The Universe in Ultraviolet Wavelength: The First Two Years of IUE, ed. R. D. CHAPMAN (NASA).
GAUSTAD, J. E., 1963, Astrophys. J., **138**, 1050.
GEHRZ, R. D. et al., 1982, Astroph. Journal, **254**, No 2, part 1, 550.
GIULI, R. T., 1968a, Icarus **8**, 301–323.
GIULI, R. T., 1968b, Icarus **9**, 186–190.
GLASSGOLD, A. E., LANGER, W. D., 1976, Astrophys. J., **204**, 403.
GÖTZ, W., 1984, Zum Charakter und Verhalten zirkumstellarer Hüllen, Veröff. Sternw. Sonneberg, **10**, H. 1.
GOLDREICH, P., WARD, W., 1973, Astrophys. J. **183**, 1051–1061.
GOLDSMITH, D. W. et al., 1969, Astrophys. J., **158**, 173.
GONDOLATSCH, F., 1981, Landolt-Börnstein. First Edition: Group VI; 2nd Vol., Springer-Verlag, Berlin, Heidelberg and New York, 129–148.
GOSWAMI, J. N., LAL, D., 1979, Icarus, **40**, 510.
GOSWAMI, J. N., LAL, D., WILKENING, L. L., 1984, Space Science Reviews **37**, 111.
GOSWAMI, J. N., MC DOUGALL, J. D., 1983, J. Geophys. Res., **88**, Suppl., A755.
GOTTFRIED, R., 1979, Über die Frühzeit des Sonnensystems und einige Phasen der Planetenentwicklung (unpublished).
GOTTFRIED, R., 1987, to be published in Gerl. Beitr. Geophys. **96** (1987).
GRAY, C. M., COMPSTON, W., 1974, Nature **251**, 495.
GRAY, C. M., PAPANASTASSIOU, D. A., WASSERBURG, G. J., 1973. Icarus, **20**, 213.
GREENBERG, J. M., 1984, Proc. Meeting "Asteroids, Comets, Meteors", Reprocentralen HSC Uppsala universitet, 259.
GROSSMAN, L., 1972, Geochim. Cosmochim. Acta, **36**, 597.
GROSSMAN, L., LARIMER, J. W., 1974, Rev. Geophys. and Space Phys., **12**, 71.
GROSSMAN, J. N., WASSON, J. T., 1983, in: Chondrules and their Origin, ed e. a. KING Lunar and Planetary Institute, Houston, 88.
HAHNE, K., 1983, Die chemische Zusammensetzung von Chondriten (unpublished).
HANKS, T., ANDERSON, D., 1969, Phys. Earth Planet. Inter. **2**, 19–29.
HANKS, T., ANDERSON, D., 1972, Phys. Earth Planet. Inter. **5**, 409–425.
HAWKE, P. S., WALLACE, W. V., BURGESS, T. J., DUERRE, D. E., HUEBEL, J. G., KEELER, R. N., KLAPPER, H., 1978, Phys. Rev. Lett. **41**, 994.

Hayashi, C., 1966, Ann. Rev. Astron. Astroph., **4**, 171.
Hayashi, C., 1970, Ann. Rev. Astron. Astrophys., **4**, 171.
Hayashi, C. et al., 1970a, Progress of Theoretical Physics, **44**, 1580.
Hayashi, C., Nakazawa, K., Adachi, I., 1977, Publ. Astron. Soc. Japan **29**, 163–196.
Hayashi, C., 1981, Suppl. of the Progress of Theoretical Physics, No. 70, 35.
Hayaski, L., Nakazawa, K., Mizuno, H., 1980, Earth Planet. Sci. Lett. **50**, 197.
Heiles, C., 1971, Ann. Rev. Astr. Astroph. **9**, 293.
Herbig, G. H., 1960, Astrophys. J. Suppl., **4**, 337.
Herbig, G. H., 1962, Advances in Astron. Astrophys., **1**, 47.
Herbig, G. H., 1972, The Astrophys. J. **174**, 401.
Herbig, G. H., 1977, Astrophys. J., **214**, 747.
Hewins, R. H., 1983, in: Chondrules and their Origin, ed. E. A. King, Lunar and Planetary Institute, Houston, 122.
Heymann, D., Dziczkaniec, M., 1976, Science **191**, 79.
Hide, R., 1982, Phil. Trans. R. Soc. Lond. A **306**, 223.
Hoerner, S. von, 1975, Lecture Notes in Physics, **42**, 53.
Horedt, G. P., 1980a, Phys. Earth Planet. Inter. **21**, 22.
Horedt, G. P., 1980b, Formation of the planets, Max-Planck-Inst. Phys. Astrophys. MPI-PAE/Astro 245.
Hubbard, W. B., 1972, Phys. Earth Planet. Inter. **6**, 65.
Hubbard, W. B., 1980, Rev. Geophys. Space Phys. **18**, 1.
Hubbard, W. B., Mac Farlane, J. J., 1980, J. Geophys. Res. **85**, 225.
Hudson, H. S., 1978, Solar Physics **57**, 237.
Hunter, C., 1965, M. N. R. A. S. **129**, 321.
Hunter, J. H., 1969, Mon. Not. Roy. Astr. Soc., **142**, 473.
Hutcheson, I. D., Steele, I. H., Smith, J. V., Clayton, R. N., 1978, Proc. Lun. Planet. Sci. Conf. 9th, 1345.
Hutchins, J. B., 1976, Astrophys. J., **205**, 103.
Hutchison, R., Bevan, A. W. R., 1983, in: Condrules and their Origin, ed. E. A. King, Lunar and Planetary Institute, Houston, 162.
— IAU Circ.: 1986, Nr. 4164.
— IAU Circ.: 1986, Nr. 4168.
Israel, F. P., Felli, M., 1978, Astron. Astrophys. **63**, 325.
Jeffery, P. M., Anders, E., 1970, Geochim. Cosmochim. Acta, **34**, 1175.
Jessberger, E. K., et al., 1988, Aspects of the Major Element Composition of Halley's Dust, submitted to Nature.
Johnston, D. H., Toksöz, M. N., 1977, Icarus **32**, 73.
Jones, D. F., Herbig, G. H., 1979, Astronom. J., **84**, 1872.
Joy, A. H., 1945, Astrophys. J., **102**, 168.
Jura, M., 1976, Astron. J., **81**, 178.
Kagan, Ju., Maksimov, L. A., Trenin, A. E., 1978, JETP **75**, 1729.
Kaiser, T., Kelly, W. R., Wasserburg, G. W., 1980, Geophys. Res. Lett. **7**, 271.
Kant, I., 1755, Allgemeine Naturgeschichte und Theorie des Himmels, Reprod. by Aufbau Verlag, Berlin, 1955.
Kaula, W. M., 1979, J. Geophys. Res. **84**, 999.
Kaula, W. M., 1980, Geol. Assoc. Can. Spec. Pap. **20**, pp. 25–34.
Kelly, W. R., Wasserburg, G. J., 1978, Geophys. Res. Lett., **5**, L 1079.

KENNEDY, G. C., HIGGINS, G. H., 1973, Moon **7**, 14.

KEONDJAN, V. P., MONIN, A. S., 1977, Tectonophysics **41**, 227.

KIESL, W., 1979, Kosmochemie, Springer-Verlag, Wien, 152.

KING, E. A., 1982, J. Geophys. Res., **87**, A429.

KIRSTEN, T., 1978, in: The Origin of the Solar System, ed. S. F. DERMOTT, Wiley, New York.

KISSEL, J., 1986, The Giotto Particulate Impact Analyse ESA SP-1077, 67—84.

KRUEGER, F. R. and KISSEL, J., 1984, Experimental investigation on ion emission with dust impact on solid surfaces, ESA SP-224, 43.

KU, W. H.-M., CHANAN, G. A., 1979, The Astrophys. J. **234**, L59.

KURAT, G., 1970, Earth Planet. Sci. Lett. **9**, 225.

KURAT, G., HOINKES, G., FREDRIKSSON, K., 1975, Earth Planet. Sci. Lett., **26**, 140.

KUSAKA, T., et al., 1970, Progr. Theoretical Physics, **44**, 1580.

KUZMIN, P., 1985, Private communication.

LANDAU, L. D., LIFSHITS, E. M., 1971, Statistische Physik, Akademie-Verlag, Berlin, 527 pp.

LANDAU, L. D., LIFSHITS, E. M., 1978, Hydrodynamik, Akademie-Verlag, Berlin, 618 pp.

— LANDOLDT-BÖRNSTEIN: 1981, 1. Edition, Vol. II, Springer, New York, 129.

LANGE, M. A., AHRENS, T. J., 1984, Earth Planet. Sci. Lett. **71**, 111.

LAPLACE, P. S., 1796, Exposition du systeme du monde, Paris.

LARIMER, J. W., 1975, Geochim. Cosmochim. Acta, **32**, 1187.

LARIMER, J. W., ANDERS, E., 1970, Geochim. Cosmochim. Acta, **34**, 367.

LARIMER, J. W., ANDERS, E., 1976, Geochim. Cosmochim. Acta **31**, 1239.

LARSON, R. B., 1969, Mon. Not. Roy. Astr. Soc., **145**, 271 and 297.

LARSON, R. B., 1972, Mon. Not. Roy. Astr. Soc., **156**, 437 and **157**, 121.

LARSON, R. B., 1973, Fund. of Cosm. Phys., Vol. 1, 1.

LARSON, R. B., 1977, in: Star Formation, IAU symposium No 76, ed. T. DE JONG and A. MAEDER, Reidel Publ. Comp., Dordrecht, 249.

LARSON, R. B., 1978, in: The Origin of the Solar System, ed. S. F. DERMOTT, Wiley, New York.

LATTIMER, J. M., 1982, in: Formation of Planetary System, ed. A. BRAHIC, Ceopadus Press.

LATTIMER, J. M., GROSSMAN, L., 1978, Moon and Planets, **19**, 169.

LATTIMER, J. M., SCHRAMM, D. N., GROSSMANN, L., 1978, Ap. J. **219**, 230.

LATTIMER, J. M., SCHRAMM, D. N., GROSSMAN, L., 1978, Astrophys. J., **219**, 230.

LAVRUKHINA, A. K., 1980, Nucleonika, **25**, 11—12, 1495.

LEE, T., PAPANASTASSIOU, D. A., 1974, Geophys. Res. Lett. **1**, 225.

LEE, T., PAPANASTASSIOU, D. A., WASSERBURG, G. J., 1976, Geophys. Res. Lett. **3**, 41.

LEE, T., PAPANASTASSIOU, D. S., WASSERBURG, G. J., 1976, Geophys. Res. Lett., **3**, 109.

LEE, T., PAPANASTASSIOU, D. A., WASSERBURG, G. J., 1978, Astrophys. J. Lett. **220**, L21.

LEE, T., 1979, Rev. Geophys. Space Phys., **17**, 1591.

LELIWA-KOPYSTYNSKÍ, J., 1984, J. Phys. C. **45**, 109.

LELIWA-KOPYSTYNSKÍ, J., TANIGUCHI, T., KONDO, K., SAWAOKA, A., 1984, Icarus **57**, 280.

LELIWA-KOPYSTYŃSKÍ, J., TEISSEYRE, R., 1984, Constitution of the Earth's interior (Physics and evolution of the Earth's interior vol. 1), PWN, Warsaw, Elsevier, Amsterdam—Oxford—New York, 368 pp.

LEQUEX, J., 1977, in: Star Formation, ed. T. DE JONG, A. MAEDER, IAU Publ.

LEVIN, B. J., 1972, Tectonophysics **13**, 7.

LEWIS, B., 1975, in: B. R. PAMPLIN (Ed.) Crystal growth, Pergamon Press, Oxford—New York—Toronto—Sydney, chap. 2, 12.

LEWIS, J. S., 1972, in: H. REEVES (Ed.) On the origin of the solar system, Centre Nacional de la Recherche Scientifique, Paris, pp. 202—205.

LEWIS, J. S., 1972, Icarus, **16**, 241.

LEWIS, R. S., SRINIVASAN, B., ANDERS, E., 1975, Science, **190**, 1251.

LITZROTH, E., 1980, Gerl. Beitr. Geophys., **89**, 350.

LITZROTH, E., 1983, Gerl. Beitr. Geophys., **92**, 509.

LJUSTICH, E. CH., 1974, Izv. Akad. Nauk SSSR, Fiz. Zemli No. 5, 17.

LODOTCHNIKOV, V. D., 1939, Zap. Mineral., O-va **64**, 207.

LOPER, D. E., 1978, Geophys. J. R. astr. Soc. **54**, 389.

LOPER, D. E., 1985, J. Geophys. Res. **90** B, 1809.

LOPER, D. E., ROBERTS, P. H., 1981, Compositional convection and the gravitationally powered dynamo, preprint.

LOREN, R. B., 1977, Astrophys. J., **218**, 716.

LORIN, J. C., CHRISTOPHE MICHEL-LEVY, M., 1978, USGS Rep., 78.

LUBIMOVA, E., 1982, Tectonophysics, **83**, 143—149.

LUCK, J. M., BIRCK, J. L., ALLEGRE, C. J., 1980, Nature, **283**, 256.

LÜST, R., 1952, Z. f. Naturforschung, A7, 87.

LYNDEN-BELL, D., PRINGLE, J., 1974, Mon. Not. Roy. Astr. Soc., **163**, 603.

LYNDEN-BELL, D., PRINGLE, J., 1974, Mon. Not. Roy. Astr. Soc., **168**, 603.

MCDOUGALL, J. D., PHINNEY, D., 1977, Proc. Lunar Sci. Conf. 8th, 293.

MCKEEGAN, K. D., SWAN, P., WALKER, R. M., WOPENKA, B., ZINNER, E., 1987, Lunar Planet. Sci. XVIII, 627.

MAHONEY, W. A., LING, J. C., WHEATON, A., JACOBSON, S., 1984, The Astroph. J., 278, 784.

MAKALKIN, A. B., 1980, Phys. Earth Planet. Inter. **22**, 302.

MALKUS, W. V. R., 1963, J. Geophys. Res. **68**, 2871.

MENDOZA, V., 1966, Astrophys. J., **143**, 1010.

MENDOZA, V., 1968, Astrophys. J., **151**, 977.

MINSTER, J. F., ALLEGRE, C. J., 1979a, Earth Planet. Sci. Lett., **42**, 333.

MINSTER, J. F., ALLEGRE, C. J., 1979b, Nature **278**, 5706, 732.

MITTELFEHLDT, D., WETHERILL, G. W., 1979, Geochim. Cosmochim. Acta., **43**, 201.

MIZUNO, H., 1980, Theor. Phys. **64**, 544.

MIZUTANI, H., MATSUI, T., TAKEMKI, K., 1972, Accretion process of the Moon, Moon, **4**, 476—489.

MÖHLMANN, D., 1971, Dissertation, Karl-Marx-Universität Leipzig.

MÖHLMANN, D., 1981, Adv. Space Res., Vol. **1** pp. 61—65.

MÖHLMANN, D., 1982, Gerl. Beitr. Geophys., **91**, 269.

MÖHLMANN, D., 1983, Gerl. Beitr. Geophys., **92**, 88.

MÖHLMANN, D., 1983a, Gerl. Beitr. Geophys., **92**, 340.

MÖHLMANN, D., 1984a, Gerl. Beitr. Geophys., **93**, 231.

MÖHLMANN, D., 1984b, Gerl. Beitr. Geophys., **93**, No. 1, p. 71.
MÖHLMANN, D., 1984c, Gerlands Beitr. Geophys., **93**, 480.
MÖHLMANN, D., 1985, Earth, Moon and Planets **33**, 201–214.
MONTMERLE, T., KOCH-MIRAMOND, L., GRINDLAY, J. E., FALGARONE, E., 1983, The Astrophys. J. **269**, 182.
— MORRISON, D., SAMZ, J., 1980, NASA-SP-439, 141.
MORTILL, G. E., VÖLK, H. J., TSCHARNUTER, W., 1985, in Protostars and Planets II, ed. D. BLACK, University of Arizona Press.
MORTILL, G. E., TSCHARNUTER, W., VÖLK, H. J., 1985a, Dynamical and Chemical Evolution of the protoplanetary nebula, MDI-H, 1985 V6.
MOUSCHOVIAS, 1979, in: Protostars and Planets, ed. T. GEHRELS, Univ. of Arizona.
MURTHY, V. R., HALL, T., 1972, Phys. Earth Planet. Int. **6**, 123,
NAKAGAWA, Y. et al., 1981, Icarus, **45**, 517.
NAKAZAWA, K., KUMORO, T., HAYASHI, C., 1983, Moon and the Planets **28**, 311–327.
NAKAZAWA, K., NAKAGAWA, Y., 1981, Origin of the Solar System, Suppl. Progr. theoret. Phys., **70**, 11–34.
NARITA, S., et al., 1970, Progr. Theor. Phys., **43**, 942.
NEECE, G. A., ROGERS, F. J., HOOVER, W. G., 1971, Journ. Compt. Phys. **7**, 621.
NEWMANN, M. J., WINKLER, K. H., 1980, Proc. Conf. Ancient Sun, ed. PEPIN, R. O., EDDY, J. A., MERRILL, R. B., p. 551.
NIEDERER, F. R., PAPANASTASSIOU, D. A., 1979, Lunar Planet. Sci. X, Lunar Planet. Inst. Houston, 913.
NIETO, M. M., 1972, The Titius-Bode law of planetary distances: Its History and Theory, Pergamon Press.
NÖLKE, F., 1930, Der Entwicklungsgang des Planetensystems, R. Dummlers Verlag, Bonn.
NORMAN, et al., 1980, Astrophys. J., **239**, 968.
NORSKOV, J. K., 1979, Phys. Rev. B **20**, 446.
NUTH, J. A., DONN, B., 1983a, J. Chem. Phys., **78**, 1618.
NUTH, J. A., DONN, B., 1983b, J. Geophys. Res., **88**, A847.
OROWAN, E., 1969, Nature **222**, 867.
OSTRIKER, J., BODENHEIMER, P., 1973, Astrophys. J., **180**, 171.
PANAGIA, N., 1973, Astron. J., **78**, 929.
PODOLAK, M., CAMERON, A. G. W., 1974, Icarus, **22**, 123.
POLYATSCHENKO, V. L., FRIDMAN, A. M., 1972, Astron. Zhurn. **49**, 157; see too Polyatschenko and Fridman Ravnovesiye i ustoitschivost gravitiruyutschych sistem, I zd. Nauka, Moscow, 1976.
RAMSAY, W. H., 1949, Mon. Not. R. astr. Soc. Geophys. Suppl. **5**, 409.
RANSFORD, G. A., 1982, Phys. Earth Planet. Inter. **29**, 209.
REEDY, R. C., ARNOLD, J. R., LAL, D., 1983, Ann. Rev. Nucl. Part. Sci. **33**, 505.
REEVES, H., 1978, Protostars and Planets, ed. T. GEHRELS, Univ. of Ariz. Press, Tucson.
RINGWOOD, A., 1975, Composition and Petrology of the Earth's mantle, McGraw-Hill, New York.
RINGWOOD, A. E., 1981, Composition and petrology of the Earth's mantle (Russ. transl.) Nedra, Moscow, 583 pp. Engl. ed. Mc Graw-Hill, New York, 1975.
RONOV, A. A., JAROSHEVSKIJ, 1976, Geochimija **12**, pp. 1763–1795 (see too Geochimija 11, 1975, pp. 1285–1309).

ROOS, M., 1974, J. Chem. Phys. **60**, 3634.
ROSS, M., 1981, Nature **292**, 435.
ROZYCZKA, M., et al., 1980, Astron. Astrophys., **81**, 347 and **83**, 118.
RÜDIGER, G., TSCHÄPE, F., 1987, to be published in Gerlands Beitr. zur Geophysik.
RUSSEL, C. T., 1980, Rev. Geophys. Space Phys. **18**, 77.
SAFRONOV, V. S., 1962, Dokl. Akad. Nauk SSSR **147**, No. 1, 64.
SAFRONOV, V. S., 1969, Evolution of the protoplanetary cloud and formation of the Earth and the planets, Nauka, Moscow (Engl. transl. NASA TT F-667, 1972).
SAFRONOV, V. S., 1972, in: H. REEVES (Ed.) On the origin of the solar system, Centre Nacional de la Recherche Scientifique, Paris, pp. 89–113.
SAFRONOV, V. S., 1972, Evolution of planetary clouds and formation of the Earth and planets, NASA TTF-677.
SAFRONOV, V., 1978, Icarus, **33**, 3–12.
SALPETER, E. E., 1979, Ann. Rev. Astron. Astrophys., **17**, 267.
SMIRNOV, W. I., 1961, Lehrgang der höheren Mathematik, Teil II, VEB Deutscher Verlag der Wissenschaften, Berlin.
— SMITH et al., 1982, Science **215**, 504.
SAVAGE, B., MATHIS, J. S., 1979, Ann. Rev. Astron. Astrophys., **17**, 73.
SCHMIDT, O. YU., 1959, A theory of the Origin of the Earth, Four Lectures (Lawrence and Wishart, London).
SCHOFIELD, N., 1981, Mon. Not. R. Astr. Soc. 197, 1031–1047.
SCHUBERT, G., CASSEN, P., YOUNG, R., 1979, Subsolidus convective cooling histories of terrestrial planets, Icarus, **38**, 192–211.
SHARPE, H. N., PELTIER, W. R., 1978, Geophys. J. R. astr. Soc. **59**, 95.
SHARPE, H., PELTIER, W., 1979, Geophys. J. R. Astr. Soc., **59**, 171–205.
SHAW, G., 1978, Phys. Earth Planet. Inter. **16**, 361.
SHAW, G., 1979, Phys. Earth Planet. Inter. **20**, 42.
SHIBATA, S., et al., 1980, The Science Reports of the Tohoku University. Eighth. Series, Vol. 1, No 3, Dec. 1980 (Sendai Astronomiaj Raportoj, N-ro 223, 1980.
SIEGFRIED, R. W., SIMMONS, G., RICHTER, D. A., HÖRZ, F., 1977, Proc. Lunar Sci. Conf. 8th, 1249.
SIEGFRIED, R. W., MC QUEEN, R. G., SIMMONS, G., 1981, J. Geophys. Res., **86**, 6205.
SKUMANICH, A., 1972, The Astrophys. J. **171**, 565.
SLATTERY, W. L., 1977, Icarus **32**, 58.
SMIRNOV, W. I., 1961, Lehrgang der höheren Mathematik, Teil II, VEB Deutscher Verlag der Wissenschaften, Berlin.
SMOLUCHOWSKI, R., 1975, Phys. Earth Planet. Inter. **20**, 247.
SMOLUCHOWSKI, R., 1983, Moon and Planets **28**, 137
SPITZER, L., 1978, Physical Processes in the Interstellar Medium, John Wiley and Sons, New York.
SPOHN, T., 1984, J. Geophys., **54**, 77–96.
STACEY, F. D., 1977, Phys. Earth Planet. Inter. **15**, 341.
STANLEY, H. E., 1973, Introduction to phase transitions and critical phenomena (Russ. transl.) Mir, Moscow, 419 pp. Engl. ed. Clarendon Press, Oxford, 1971.
STEIN, R. F., et al., 1972, Astrophys. J. Lett., **177**, L 125.
STEPHENS, J. R., BAUER, S. H., 1981, Meteoritics, **16**, 388.
STEPHENS, J. R., RUSSEL, R. W., 1979, Astrophys. J. Lett., **228**, 780.

STERN, R. A., 1984, in: Proc. of the Third Cambridge Workshop on Cool Stars, Stellar Systems and the Sun, ed. S. L. BALIUNAS and L. HARTMANN, Springer, Berlin, pp. 150.

STEVENSON, D. J., 1977, Astrophys. J. Suppl. Ser. **35**, 239.

STEVENSON, D. J., 1978, The outer planets and their satellites, in: S. F. DERMOTT (Ed.) The origin of the solar system, J. Wiley & Sons, New York, pp. 395–431.

STEVENSON, D. J., 1980, Science **208**, 746.

STEVENSON, D. J., 1981, Science **214**, 611.

STEVENSON, D. J., 1982a, Ann. Rev. Earth Planet. Sci. **10**, 257.

STEVENSON, D. J., 1982b, Planet. Space Sci. **30**, 755.

STEVENSON, D. J., SALPETER, E. E., 1976, in: T. GEHRELS (Ed.) Jupiter, University of Arizona Press, Tucson, Ariz., 85.

STEVENSON, D. J., SALPETER, E. E., 1977, Astrophys. J. Suppl. Ser. **35**, 239.

STEWARD, G. R., LIN, D., BODENHEIMER, P., 1984, in Planetary Rings, eds. R. GREENBERG and A. BRAHIC, pp. 447–512, University of Arizona Press, Tucson.

STILLER, H. FRANCK, S., 1980, Phys. Earth Planet. Inter. **22**, 216.

STILLER, H., MÖHLMANN, D., 1981, Spectrum (Akad. d. Wiss. d. DDR, Berlin), **12**, No. 10, 6.

STILLER, H., SEIPOLD, U., VOLLSTÄDT, H., 1974, Fiz. svoistva goru. porod i miner. pri vys. temp., Mat. Vsesojusu. Sovésc. **4**, 198.

STILLER, H., SEIPOLD, U., VOLLSTÄDT, H., 1975, Geodaet. geophys. Veröff. Berlin, R III **36**, 51.

STILLER, H., TREDER, H. J., MÖHLMANN, D., 1981, Adv. Space Res. **1**, 21.

STILLER, H., VOLLSTÄDT, H., FRANCK, S., 1981, State of matter inside the Earth and some implications for geodynamics, contribution to the 4th International Symposium and Physics of the Earth, Karl-Marx-Stadt.

STIX, M., 1977, J. Geophys. **43**, 695.

— STONE, E. C., MINER, E. D., 1982, Science **215**, 499.

— STONE, E. C., MINER, E. D., 1986, Science **233**, 39.

STRETT, W. B., 1974, Astrophys. J. **186**, 1107.

STROM, K. M., et al., 1972, Astrophys. J., **171**, 267 and **173**, 353.

SUZUKI, T., 1983, The system iron-enstatite-water at high pressures and temperatures-formation of iron hydride and some geophysical implications, Technical Report of ISSP, University of Tokyo, Ser. A No. 1384.

SZETO, M. K., 1983, Icarus **55**, 133.

TATSUMOTO, M., KNIGHT, R. J., ALLEGRE, C. J., 1973, Science, **180**, 1279.

TATSUMOTO, M., UNRUH, D. M., DESBOROUGH, G. A., 1976, Geochim. Cosmochim. Acta **40**, 617.

TAYLOR, S., 1979, Structure and evolution of the Moon, Nature, **281**, 105–110.

TERA, F., WASSERBURG, G. J., 1974, Earth Planet. Sci. Lett. **14**, 281.

THIEMENS, M. H., HEIDENREICH, J. E., 1983a, Science, **219**, 1073.

THIEMENS, M. H., HEIDENREICH, J. E., 1983b, J. of Chem. Phys., **78**, 892.

TOKSÖZ, M. N., HSUI, A. T., JOHNSTON, D. H., 1978, Moon and Planets **18**, 281.

TOKSÖZ, M., SOLOMON, S., MINEAR, J., JOHUSTON, P., 1972, Moon, **4**, 190–213.

TOOMRE, A., 1964, Astrophys. Journal, **139**, 1217.

TORBETT, M., GREENBERG, R., SMOLUCHOWSKI, R., 1982, Icarus **49**, 313–326.

TORBETT, M., SMOLUCHOWSKI, R., 1980, Nature **286**, 237.

TOZER, D. C., 1965, Geophys. J. R. astr. Soc. **9**, 95.

TOZER, D. C., 1978, Terrestrial planet evolution and the observational of their formation, in: S. F. DERMOTT (Ed.) The origin of the solar system, J. Wiley & Sons, New York, pp. 433–462.

TRIMBLE, V., 1975, Rev. Mod. Phys., **47**, 877.

TRUBITSYN, V. P., 1966, Fiz. tverd. tela **8**, 862.

TSCHARNUTER, W. M., 1985, private communication, and in Birth and infancy of stars, pp. 601–643, Elsevier science Publi. B. V.

TUREKIAN, K., CLARK, S. P., 1969, Earth Planet. Sci. Lett. **6**, 346.

TUREKIAN, K. K., WEDEPOHL, K. H., 1961, Bull. Geol. Soc. Amer. **72**, pp. 172–202.

ULRICH, R. K., 1976, Astrophys. J., **210**, 377.

USSELMAN, T. M., 1975, Am. J. Sci. **275**, 291.

VAHIA, M. N., BISWAS, S., 1982, in: Composition and Origin of Cosmic Rays, ed. M. M. SHAPIRO, D. Reidel Publishing Company, Dordrecht, pp. 155.

VARDANIAN, R. A., 1964, Soobsch Bjurakan Obs., **35**, 3.

VASUNDHARE, R., 1984, Nature, **312**, 621–623.

VERHOOGEN, J., 1973, Phys. Earth Planet. Inter. **7**, 47.

VINOGRADOV, A. P., 1962, Geochimija **7**, pp. 555–571.

VITJAZEV, A. V., 1973, Izv. Akad. Nauk. SSSR, Fiz. Zemli No. 10, 86.

VITJAZEV, A. V., 1982, Izv. Akad. Nauk SSSR, Fiz. Zemli No. 6, 52.

VITJAZEV, A. V., MAJEVA, S. V., 1977, Tectonophyscis **41**, 217.

VITJAZEV, A. V., MAJEVA, S. V., 1980, Phys. Earth Planet. Inter. **22**, 296.

VÖLK, H., 1981, Mitt. Astron. Ges. **51**, 63.

VÖLK, H., 1984, Astrophys. Journ. **283**, 371–395.

VOGEL, S., KUHI, L. V., 1981, Astrophys. J., **245**, 934.

WÄSCH, R., 1983, PI **9**, No. 12, 20.

WÄSCH, R., 1986, An approach to the nature of the silicatic component of the comet Halley, submitted to Proc. 20th ESLAB Symposium Exploration of Halley's Comet, Heidelberg, FRG.

WALKER, M. F., 1972, Astrophys. J., **175**, 89.

WALTER, F. M., 1984, The Astrophys. J. **284**, 194.

WALTER, F. M., KUHI, L. V., 1981, The Astrophys. J., **250**, 253.

WARD, W. R., 1981, Icarus, **47**, 234–264.

WARK, D. A., 1979, Astrophys. and Space Sci., **65**, 275.

WASSERBURG, G., PAPANASTASSIOU, D. A., LEE, T., 1980, in: Proc. of the Internat. School of Phys. „E. Fermi", Course LXXIII, Early Solar System Process, pp. 144, Amsterdam.

WASSERBURG, G. J., PAPANASTASSIOU, D. A., 1982, in: Essays in Nuclear Astrophys. ed. W. F. FOWLER, Cambridge, pp. 77.

WASSERBURG, G. J., TERA, F., PAPANASTASSIOU, D. A., HUNEKE, J. C., 1977, Earth Planet. Sci. Lett., **35**, 299.

WEDEPOHL, K. H., 1969, Fortschr. Mineralog. **46**, 2, pp. 145–174 and Handbook Geochemistry Bd. 1, Springer-Verlag Berlin, pp. 227–271.

WEGENER, A., 1980, Die Entstehung der Kontinente und Ozeane, (Nachdruck), Akademie-Verlag Berlin.

WEIDENSCHILLING, S. J., 1974, A model for accretion of the terrestrial planets. Icarus **22**, 426–435.

WEIDENSCHILLING, S. J., 1977, Astrophys. Space Sci., **51**, 153.
WEIDENSCHILLING, S. J., 1977a, Mon. Not. Roy. Astr. Soc., **180**, 57.
WEIDENSCHILLING, S., 1980, Dust to planetesimals. Settling and coagulation in the Solar Nebula, Icarus, **44**, 172–189.
WEIDENSCHILLING, S. J., DAVIS, D. R., 1985, Icarus, **62**, 16–29.
WILDT, R., 1972, Phys. Earth Planet. Inter. **6**, 1.
WILLIAM, I. P., CRAMPIN, D. J., 1971, Mon. Not. Roy. Astr. Soc., **152**, 261.
WINKLER, K.-H., NEWMAN, M. J., 1980, Astrophys. J., **236**, 201 and **238**, 311.
WOOD, D., 1981, MNRAS, **194**, 201.
WOOD, J. A., 1983, Formation of Chondrules and CAI's from interstellar grains accreting to the solar nebula, Mem. Natl. Inst. Polar Res. Spec. Issue, Tokyo.
WORDEN, S. P., SCHNEEBERGER, T. J., KUHN, J. R., AFRICANO, J. L., 1981, The Astrophys. J. **244**, 520.
YAMAGITA, S., GENSHO, R., 1977, Astrophys. Space Sci., 46, L1.
YOMOGIDA, K., MATSUI, T., 1982, Proc. 7th Symposium on Antarctic Meteorites, Natl. Inst. of Polar Res., Tokyo Special Issue Nr. 25.
YOMOGIDA, K., MATSUI, T., 1983, J. Geophys. Res., **88**, No. 111, 9513.
YOMOGIDA, K., MATSUI, T., 1984, Earth Planet. Sci. Lett., **68**, 34.
YORKE, H. W., KRÜGEL, E., 1977, Astron. Astrophys., **54**, 183.
YOUNG, R. E., SCHUBERT, G. 1974, Geophys. Res. Lett. **1**, 157–160.
ZAPOLSKI, H. S., SALPETER, E. E., 1969, Astroph. j. **158**, 809.
ZHARKOV, V. N., 1983, Moon and Planets **29**, 139.
ZHARKOV, V. N., TRUBITSYN, V. P., 1980, Phys. of the planetary interiors, Nauka, Moscow, 448 pp. Engl. transl. Patchart, Tucson, Ariz., 1980.
ZUCKERMANN, B., PALMER, P., 1974, Ann. Rev. Astron. Astroph., **12**, 279.

Index

Bei Fragen zur Produktsicherheit wenden Sie sich bitte an:
If you have any questions regarding product safety,
please contact:

Walter de Gruyter GmbH
Genthiner Straße 13
10785 Berlin
productsafety@degruyterbrill.com